Handbook of Electronic Components and Circuits

John D. Lenk
Consulting Technical Writer

PRENTICE-HALL, INC., Englewood Cliffs, New Jersey

Library of Congress Cataloging in Publication Data

LENK, JOHN D.
Handbook of electronic components and circuits

1. Field effect transistors and unijunction transistors
I. Title
QA78.561234 002.7′53 74-678913′
ISBN 0-13-377283-7

10 9 8 7 6 5 4 3 2 1

Printed in the United States of America

PRENTICE-HALL INTERNATIONAL, INC., *London*
PRENTICE-HALL OF AUSTRALIA, PTY., LTD., *Sydney*
PRENTICE-HALL OF CANADA, LTD., *Toronto*
PRENTICE-HALL OF INDIA PRIVATE LIMITED, *New Delhi*
PRENTICE-HALL OF JAPAN, INC., *Tokyo*

Preface

This book is a companion to the author's *Handbook of Simplified Solid-State Circuit Design.* The solid-state circuit design book concentrates on diodes, conventional bipolar (two-junction) transistors, and IC operational amplifiers, to provide approximations or rules of thumb for the selection of circuit components. *Handook of Electronic Circuits and Components* provides the same type of coverage for unijunction transistors (UJT) and field effect transistors (FET). Both books assume a specific design goal and a given set of conditions. Both books concentrate on simple, practical approaches to circuit design, not on circuit analysis. Theory is kept to a minimum.

With any solid-state circuit, it is possible to apply certain guidelines for the selection of component values. These rules can then be stated in basic equations, requiring only simple arithmetic for their solution.

The component values will depend upon the solid-state device characteristics, available power sources, the desired performance (voltage output, stability, etc.), and external circuit conditions (input/output impedance match, input signal amplitude, etc.). Solid-state device characteristics are to be found in manufacturers' data. Circuit characteristics can then be determined, based on reasonable expectation of the device characteristics. Often, the final circuit is a result of many tradeoffs between desired performance and available characteristics. This handbook discusses the problem of tradeoffs from a simplified, practical standpoint. Not all possible uses for UJT and FET devices are covered; instead, the book concentrates on those uses for which the devices are best suited.

It is assumed that the reader is already familiar with solid-state basics at a level equivalent to that found in the author's *Practical Semiconductor Databook for Electronic Engineers and Technicians.* It is further assumed that the reader has a knowledge and understanding of basic electronic equations (resonance, circuit *Q*, time-frequency relationships, and the like) such as those tabulated in the author's *Databook for Electronic Technicians and Engineers.* It is also recommended that readers study the companion book, *Handbook of Simplified Solid-State Circuit Design.* However, *no direct reference to any of these books is required to understand and use this book.*

Since this book does not require advanced math or theoretical study, it is ideal for the experimenter. In addition, the book is well suited to schools where the basic teaching approach is circuit analysis and a great desire exists for practical design.

JOHN D. LENK

How to Use This Book

Once you have read this introduction, you may go directly to the chapter and section that describes the design procedures for a particular circuit or device. Use the Contents or Index to locate the first page of the chapter or section. Unijunction transistors (UJT) are covered in Chapter 1. Field effect transistors (FET) are discussed in Chapter 2. In turn, a separate section is assigned to each type of circuit within the related chapter. The sections are either complete within themselves, or make reference to another specific section (by section number).

The same format or pattern is used in each section where practical:

First, a working schematic is presented for the circuit, together with a brief description of the operational theory. The working schematic also includes the operational characteristics of the circuit (in equation form) as well as guidelines for selection of the circuit values.

Next, design considerations such as desired performance, use with external circuits, and available (or required) power sources are discussed. Each major design factor (supply voltage, amplification, operating frequency, device characteristics, and the like) is covered. This is followed by reference to the equations and procedures for determining component values that will produce the desired results.

Finally, a design example is given, where such examples are essential to understand the circuit from the practical viewpoint. A specific design problem is stated. The value of each circuit component is calculated in step-by-step procedures, using the guidelines established in the design considerations.

Dedicated to my wife, Irene.
This is her second sea-house book.

Contents

CHAPTER 1

Unijunction Transistor Circuits

1-1 Basic Unijunction Relaxation Oscillator

The relaxation oscillator is the basic building block in most unijunction transistor (UJT) timer and oscillator circuits. Figure 1-1 shows the basic circuit, together with some typical waveforms. Operation of the circuit is as follows.

When power is applied, the capacitor C_E charges exponentially through the resistor R_E until voltage on the capacitor equals the emitter firing voltage (V_p). At this voltage, the emitter Base 1 junction becomes forward-biased, and the emitter goes into the negative resistance region of its characteristic curve. The capacitor C_E discharges through the emitter, and a positive-going pulse will be available at Base 1. This pulse is shown in Fig. 1-1b. The circuit values used to produce the pulses are also shown in Fig. 1-1.

Prior to firing, a current I_{B2} is flowing from Base 2 to Base 1. When emitter current starts to flow, this current will increase since the resistance from Base 2 to ground is decreasing. A negative-going voltage pulse will appear therefore at Base 2. This waveform is shown in Fig. 1-1c.

When voltage at the emitter has decreased to peak emitter voltage V_p, (a voltage approximately equal to the valley voltage (V_v) when R_1 is purely resistive), the UJT will turn off if R_E meets certain conditions. C_E will start

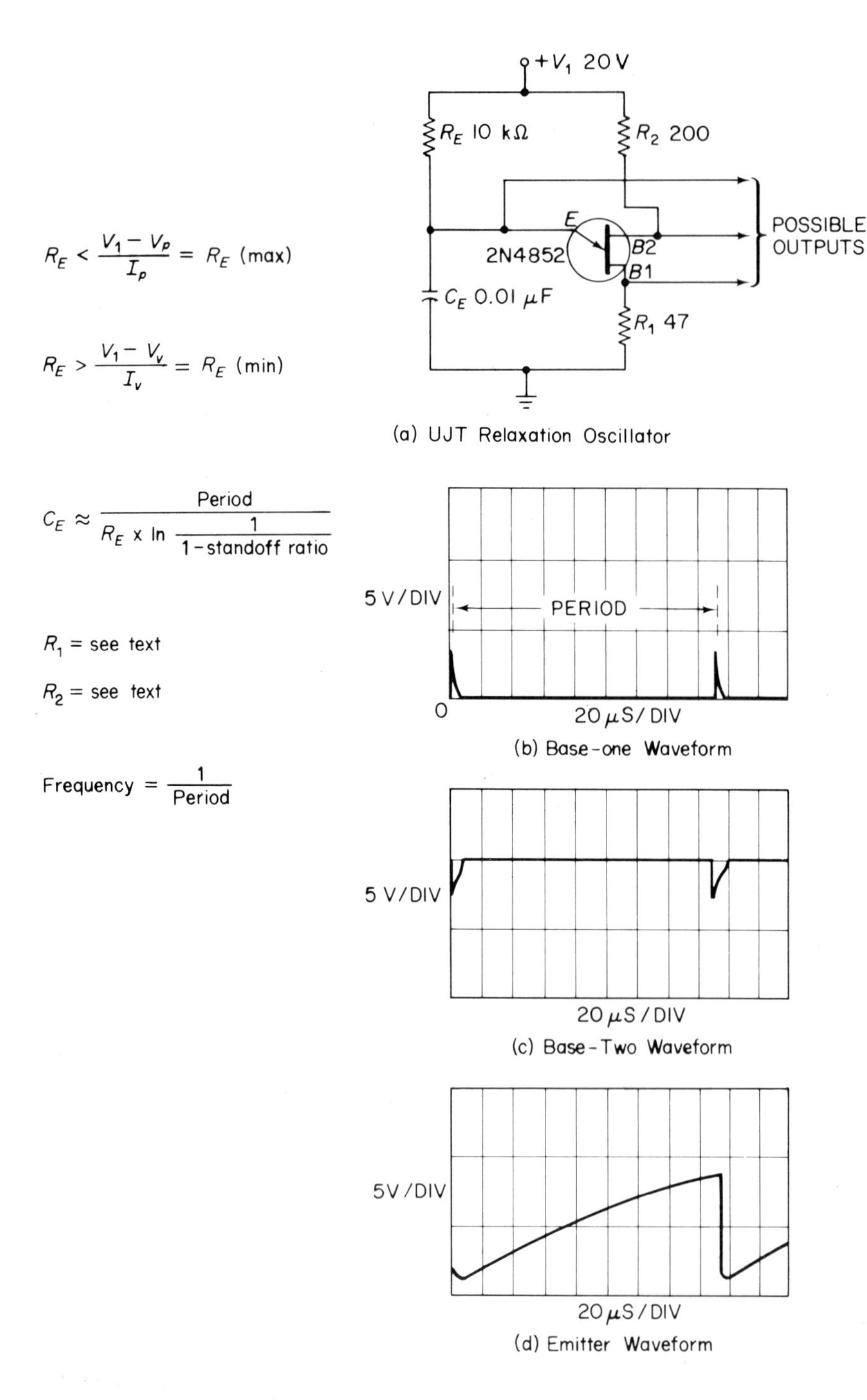

Fig. 1-1 Basic UJT relaxation oscillator (Courtesy of Motorola Inc., Semiconductor Products Division)

to charge up again, and the cycle repeats. The waveform that appears at the emitter is shown in Fig. 1-1d.

1-1-1 Design Considerations

The period of oscillation (and thus the frequency) is determined primarily by the values of R_E and C_E, as shown by the equations of Fig. 1-1. It is most practical to start design by selecting a trial value for R_E, rather than for C_E, because R_E must meet certain conditions for the oscillator to operate. If R_E is too large, the UJT will never fire; if R_E is too small, the UJT will not turn OFF.

These conditions can best be explained by means of the emitter characteristic curve in Fig. 1-2. (This curve is not drawn to scale in order to have more detail.)

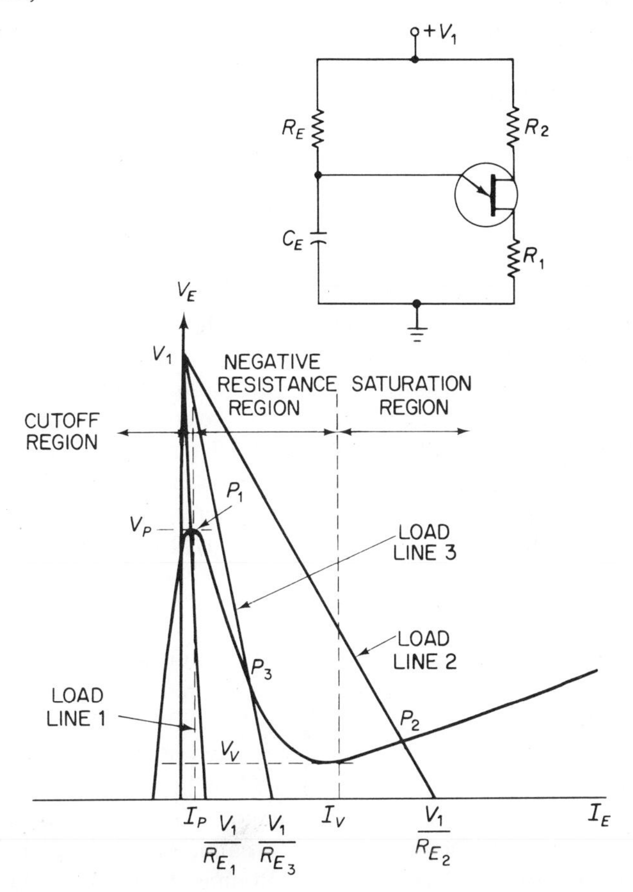

Fig. 1-2 UJT emitter characteristic load lines (Courtesy of Motorola Inc., Semiconductor Products Division)

SELECTING EMITTER RESISTOR R_E

The emitter capacitor C_E will charge until the emitter voltage is equal to V_p. At this point on the characteristic curve, peak-point emitter current I_p will be flowing and, in order to fire the UJT, the value of R_E must be small enough to allow a current somewhat larger than I_p to flow. R_E must, therefore, meet the following requirement:

$$R_E < \frac{V_1 - V_p}{I_p} = R_{E(\max)} \tag{1-1}$$

where V_1 is the applied bias voltage.

Referring to Fig. 1-2, a load line intersecting the characteristic curve in the *cutoff region*, as illustrated by load line 1, would keep the UJT from ever firing. Therefore, R_E must be small enough to keep the UJT in the *negative resistance region.*

By keeping R_E smaller than $R_{E(\max)}$, the UJT will turn ON, and C_E will discharge through the emitter. However, if R_E is too small, and an emitter current larger than the valley current I_V flows, the UJT will not turn OFF. Under these conditions, the UJT will operate in the *saturation region*, as illustrated by load line 2 in Fig. 1-2. The minimum R_E that can be used in order to assure oscillation is set by the following condition:

$$R_E > \frac{V_1 - V_V}{I_v} = R_{E(\min)} \tag{1-2}$$

where V_1 is the applied bias voltage.

An emitter resistance R_E selected to meet the requirements in Eqs. (1-1) and (1-2) will result in a load line that intersects the characteristic curve somewhere in the *negative resistance region.* This is illustrated by load line 3 in Fig. 1-2. However, in a practical UJT circuit, the emitter voltage variation in the neighborhood of the valley point is small. Thus, in order to assure turn-off, the value of R_E should be *two to three times larger than* $R_{E(\min)}$. This is best understood by a study of the following paragraphs which summarize the dynamic operating conditions of a typical UJT oscillator.

DYNAMIC OPERATING PATHS OF UJT OSCILLATOR

Figure 1-3 shows the basic UJT oscillator circuit with typical component values and the corresponding emitter voltage-current operating curves, both static and dynamic.

As shown on the curves, C_E will start to charge from point A. At point B, where the voltage of C_E equals V_p, the UJT will fire, and the characteristic curve goes into the negative resistance region. The voltage on C_E cannot change instantaneously, however, so the dynamic operating path will move from point B to point C. This is the turn-on time of the UJT and is typically less than 1 μS.

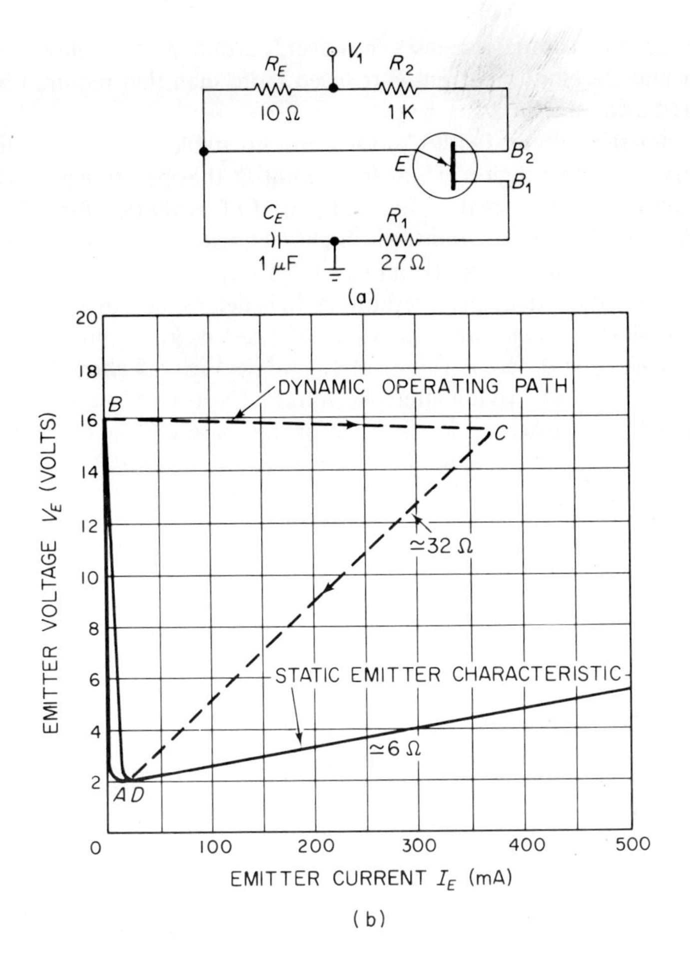

Fig. 1-3 Typical UJT oscillator emitter voltage-current operating curves (Courtesy of Motorola Inc., Semiconductor Products Division)

From point C, the operating path follows an essentially straight line to point, D, which is about equal to the valley point. The straight line has a slope of approximately 32 volts/ampere, or 32 ohms, and is composed of R_1 and the UJT emitter to Base-1 saturation resistance. There is not enough emitter current available to sustain operation at point D, and the operating path tries to follow the characteristic curve to the point where it is intersected by the load line determined by R_E (similar to load line 3 in Fig. 1-2).

If the emitter circuit had no capacitor C_E (an impractical circuit), this intersection point would be a stable operating point. To reach this point, however, the emitter voltage must increase. The resistance from emitter to Base-1 will also increase, and the emitter current will decrease somewhat.

When the emitter voltage increases, however, current starts to flow into the capacitor and the emitter current is reduced more than that required by the characteristic curve.

In a practical circuit (with C_E) there are no stable operating points in the negative resistance region. Thus, from point D, the operating path goes to point A, and the cycle repeats. The total period of oscillation (and thus the frequency) is set by turn-on and turn-off characteristics of the UJT, plus the charge time of C_E through R_E (from point A to B).

The shape of the dynamic operating path is determined by the capacitor C_E, the interbase voltage, and the value of resistor R_1. Figure 1-4 shows operating paths for different values of C_E, while Fig. 1-5 shows operating paths for fixed C_E but varying interbase voltage. (Note that interbase voltage is shown as V_{B2B1}. However, V_{BB} is used for interbase voltage on some UJT data sheets.) No operating path curves are given for various R_1 values. This

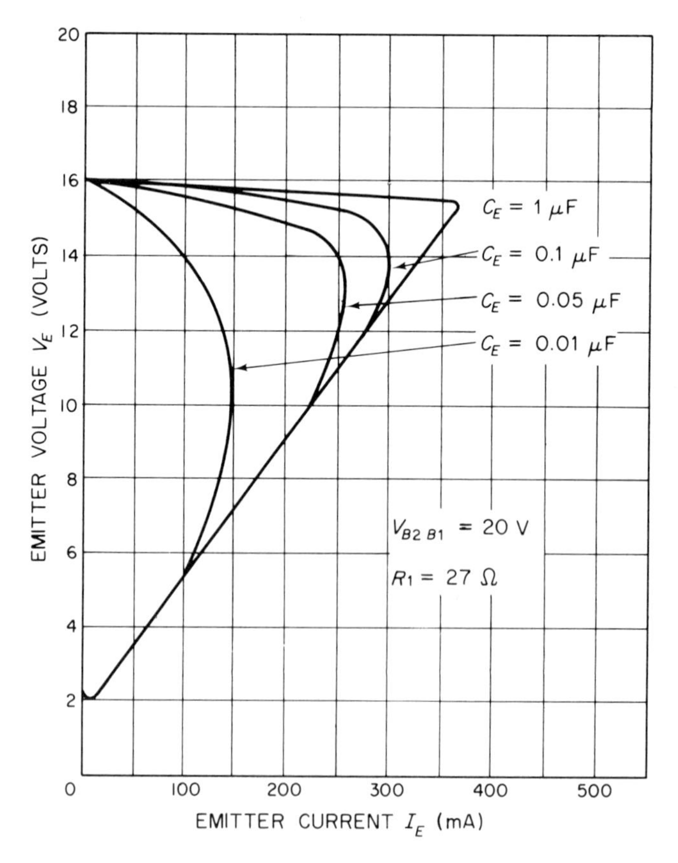

Fig. 1-4 Dynamic operating path versus emitter capacitance (Courtesy of Motorola Inc., Semiconductor Products Division)

is because the value of R_1 is generally selected on the basis of required minimum output pulse voltage, or on minimum fixed level voltage. The factors affecting the value of R_1 are discussed later in this section.

The curves of Figs. 1-4 and 1-5 bring up some important tradeoffs. With all other factors equal, the maximum available output voltage is set by the maximum emitter current I_E. As shown in Fig. 1-4, I_E can be increased when C_E is increased. Likewise, as shown in Fig. 1-5, I_E can be increased when interbase voltage (V_{B2B1}) is increased. If either C_E or interbase voltage is increased, the period of oscillation is increased and the frequency is decreased. Thus, a higher output voltage from the UJT oscillator must be traded for lower operating frequency.

Keep in mind that the V_{B2B1} voltage shown in Figs. 1-4 and 1-5 is not necessarily the V_1 (source or supply) voltage. Under static conditions, there can be some current flowing through Base-1 and Base-2. This relationship

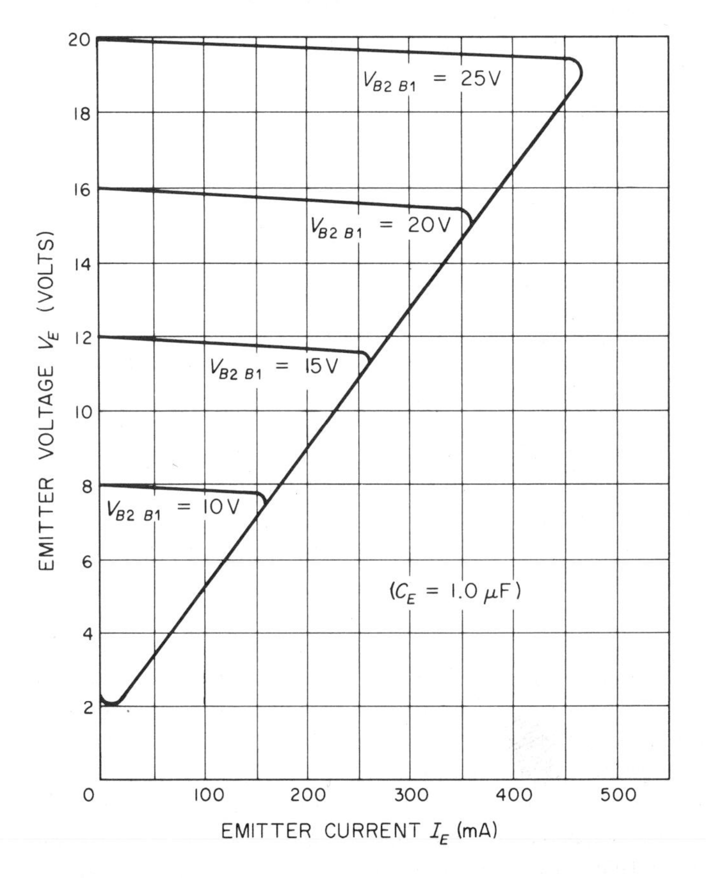

Fig. 1-5 Dynamic operating path versus interbase voltage (Courtesy of Motorola Inc., Semiconductor Products Division)

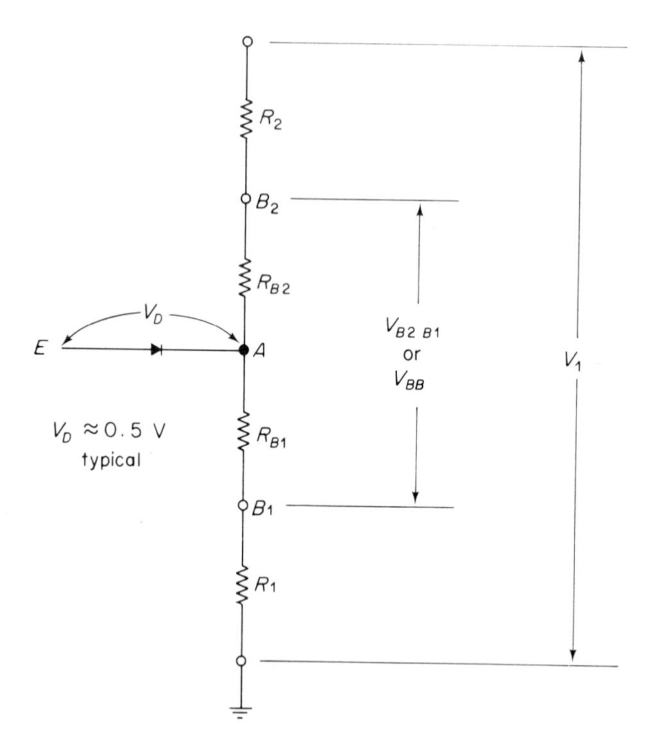

Fig. 1-6 Electrical equivalent circuit for the UJT with the external compensating resistor R_2 and output resistor R_1. The equivalent circuit is valid in the cutoff region only. (Courtesy of Motorola Inc., Semiconductor Products Division)

is shown in Fig. 1-6. The amount of current is set by the interbase resistance (r_{BB}), the value of R_2, and (to a lesser extent) by R_1 (since R_1 is generally quite small in relation to R_2 and r_{BB}). For example, assume that r_{BB} is 7 kΩ, V_1 is 20 V, and R_2 is 1 kΩ; approximately 2.5 mA will flow. This will produce an approximate drop of 2.5 V across R_2, making the interbase voltage about 17.5 V. (Note that interbase resistance is shown as r_{BB} on some data sheets, and R_{BB} on others.)

When an inductive load, like a relay coil, for example, is substituted for R_1, the dynamic operating path will be somewhat different. Figure 1-7a shows a relaxation oscillator having a pulse transformer instead of R_1, and Fig. 1-7b shows the resulting dynamic operating path. An important difference here is that the emitter no longer ceases to conduct when valley voltage (typically about 2 V) is reached, but continues down to less than 0.5 V before turning OFF. Figure 1-7c shows how the turn-off voltage is dependent on the bias voltage.

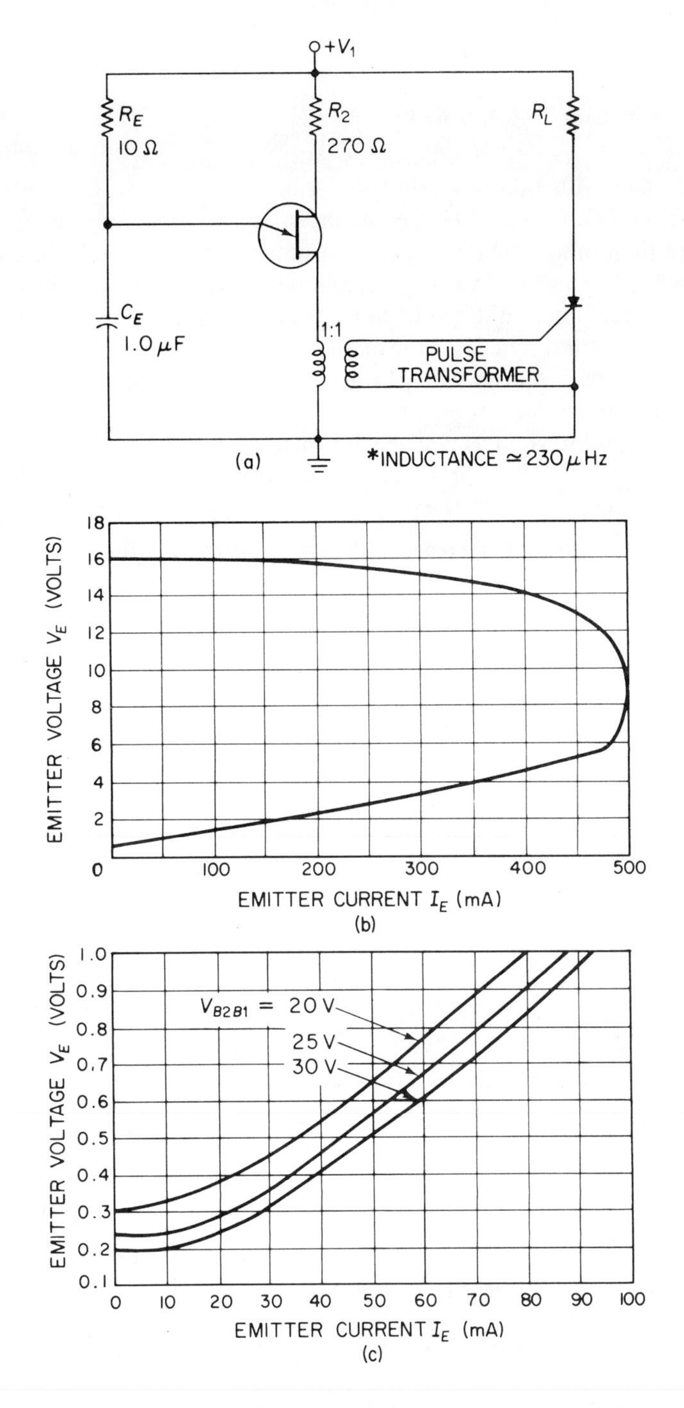

Fig. 1-7 UJT relaxation oscillator with inductive load (Courtesy of Motorola Inc., Semiconductor Products Division)

SELECTING BASE 2 RESISTOR R_2

The primary function of R_2 (in Fig. 1-1 or 1-3) is to provide temperature compensation. Practically all UJT characteristics are temperature dependent, some more than others. The interbase resistance and emitter reverse current increase, while the peak and valley voltage (and current), the intrinsic standoff ratio (η), and the junction diode drop decrease with increasing temperature. Generally, oscillator frequency is the main factor affected by these temperature variations. However, output voltage can also be affected and is of some concern in many applications.

The peak point voltage is given by this equation:

$$V_p = V_D + \eta V_{B2B1} \tag{1-3}$$

Since both V_D and η decrease with temperature, V_p will also decrease. This is, of course, a very undesirable condition in many applications, particularly in timers and oscillator circuits. It has been found, however, that the change in V_p can be compensated for by adding a resistor R_2 in series with the Base 2 terminal.

If R_2 is selected properly, V_p can be made to vary less than 1% over a 50°C temperature variation. The value of R_2 can be selected by using either the equation, or the curves, of Fig. 1-8.

The curves of Fig. 1-8 show frequency variation as a function of temperature for a typical UJT. Temperature curves for several values of R_2 ranging from 250 ohms to 3 kΩ are shown, and an R_2 of approximately 1.5 kΩ can

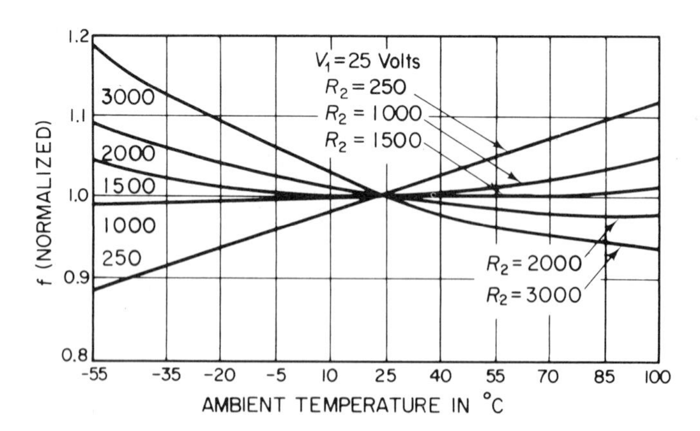

Fig. 1-8 Frequency versus temperature for a UJT relaxation oscillator. Frequency is normalized to 25°C, and R_2 is a variable parameter. (Courtesy of Motorola Inc., Semiconductor Products Division)

be seen to compensate very well from $-5°C$ to $+85°C$. A smaller resistance should be used for operation below $-5°C$.

The equation of Fig. 1-8 can be used for a somewhat more accurate "optimum" value of R_2. The equation takes into account source voltage, interbase resistance, and standoff ratio. Keep in mind that both the curves and equation of Fig. 1-8 are for trial values. Also keep in mind that an increase in R_2 value will decrease interbase voltage. As discussed, this increases frequency, but decreases output voltage (all other factors being equal).

SELECTING BASE 1 RESISTOR R_1

The primary function of R_1 (in Fig. 1-1 or 1-3) is to provide an output load for the oscillator. In some applications, R_1 is included to provide a path for the interbase current. Such a path is necessary in some circuits to prevent current from flowing through the device being driven by the UJT oscillator.

Typically, R_1 is less than 100 ohms, but could be as high as 2 or 3 kΩ in some applications.

When R_1 is selected on the basis of a current path for an external device, a maximum voltage drop is usually specified. For example, if the UJT oscillator is to trigger an SCR (from an output pulse across R_1), and the fixed SCR voltage must not exceed 50 mV (between firing cycles), the drop across R_1 must not exceed 50 mV. Assuming a 2.5-mA interbase current (from the previous example), and the 50-mV maximum voltage drop, the value of R_1 would be 20 ohms.

When R_1 is selected on the basis of output voltage, which is usually the case, a minimum output voltage is usually specified. For example, if the UJT oscillator is to trigger an SCR or other thyristor device, and the SCR requires a minimum of 3 V for the trigger, the peak drop across R_1 must be 3 V. If curves such as shown in Figs. 1-3 through 1-5 are available, it is possible to calculate an approximate value of R_1, using the peak emitter current I_E and the desired output voltage. For example, if the output is 3 V and the peak I_E is 300 mA, the value of R_1 is 10 ohms. However, the shape of the dynamic operating curve will change with a change in R_1 (and with changes in C_E).

For this reason, the curves of Fig. 1-9 are provided. In Fig. 1-9, the peak voltage across R_1 is shown as a function of C_E for various values of R_1. These are *minimum* peak output values. In practice, the peak output across R_1 will usually be 25 to 50 per cent higher. The values shown are for a supply voltage of 20 V. The peak voltage amplitude at other values of supply voltage V_1 can be obtained by multiplying the values from Fig. 1-9 by the factor shown. For example, if the supply voltage is 25 V, the peak voltage output for an R_1 of 10 ohms and a C_E of 1 μF is about 4.2 V. This is found as follows: The curve of Fig. 1-9 shows a minimum output of 3 V for an R_1 of 10 ohms and a C_E

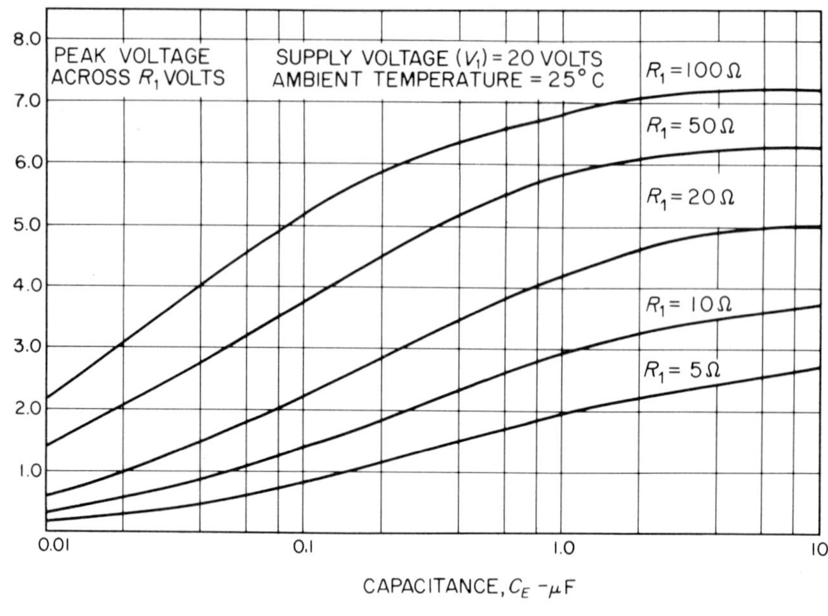

Fig. 1-9 Peak output voltage across R_1 in UJT trigger circuit—minimum values (Courtesy of General Electric Semiconductor Products Department, Syracuse, N.Y.)

of 1 μF. Using the correction factor for 25 V of approximately 1.4 (25 – 6/14), the approximate minimum output voltage is 1.4 × 3 V = 4.2 V.

Selecting Emitter Capacitor C_E

The value of capacitor C_E is determined by the desired period (or frequency) of operation. The equation for C_E shown on Fig. 1-1 is *very approximate*, since it does not take into account the ON and OFF times of the UJT, or other characteristics of the circuit such as source voltage (V_1), emitter to Base 1 voltage drop (V_D), valley voltage (V_v), standoff ratio, or interbase voltage (V_{B2B1} or, simply, V_{BB}). However, the period equation of Fig. 1-1 for C_E is accurate enough for a first trial value.

1-1-2 Design Example

Assume that the circuit of Fig. 1-1 is to provide an output of 1.5 V (minimum) across R_1 at a frequency of approximately 5.6 kHz. The available source voltage V_1 is 20 V. A test of the UJT (or data-sheet information) shows a typical valley voltage (V_v) of 2 V, and a typical valley current (I_v) of 6 mA. The standoff ratio is 0.87, and the interbase resistance (r_{BB}) is approximately 8 kΩ.

Design starts with R_E. Equation (1-2) shows that the absolute minimum value, or $R_{E(\text{min})}$, is 3 kΩ ($20 - 2 = 18$; $18/0.006 = 3$ kΩ). The first trial value of R_E should be three times $R_{E(\text{min})}$ or 9 kΩ. Use about 10 kΩ in this case. In a practical experimental circuit, all of the components should be assembled, using a 20 or 25 kΩ potentiometer for R_E. The potentiometer is set to 10 kΩ, and C_E is adjusted (or selected) for the desired frequency. With the oscillator operating at the correct frequency (or producing the correct period), R_E is adjusted to twice the selected value (or about 20 kΩ) and one-half the selected value (or about 5 kΩ).

The frequency will shift when R_E is adjusted, but the oscillator should continue to oscillate. If so, the value of R_E (10 kΩ) is such that the UJT is well within the negative resistance region, and the oscillator should continue to operate in spite of changes in supply voltage, temperature, etc. On the other hand, if the oscillator stops with small changes in R_E value, the UJT is too near the saturation region. Use an R_E value of at least one order higher. Keep in mind that an increase in R_E value will require a corresponding decrease in C_E to produce the same frequency. If R_E is increased without a change in C_E, period will increase and frequency will decrease.

With the value of R_E established at 10 kΩ, the value of C_E is selected to provide the correct frequency. The desired frequency is approximately 5.6 kHz, which requires a period of about 175 μS. With a period of 175 μS, an R_E of 10 kΩ, and a standoff ratio of 0.87, the equation of Fig. 1-1 shows that C_E should be approximately 0.01 μF.

With a value of 0.01 μF for C_E, and a required minimum of 1.5 V peak output across R_1, Fig. 1-9 shows a value of 50 ohms for R_1.

The value of R_2 is found by using the equation of Fig. 1-8. With a source voltage of 20 V, a standoff ratio of 0.87, and interbase resistance of about 8 kΩ, the value of R_2 is $0.015 \times 20 \times 0.87 \times 8000 = 2088$ ohms (use 2 kΩ).

In a practical experimental circuit, the UJT is subjected to variations in temperature, and any frequency shift is noted. (A simple test is to hold a soldering tool near, but not touching, the oscillating UJT.) There will be some frequency variation with changes in temperature no matter what value of R_2 is selected. However, the optimum value for R_2 is that which produces the least amount of frequency change with temperature variation. In practical terms, the circuit of Fig. 1-1 could operate with R_2 values of about 200 to 3000 ohms. If the circuit is to operate over a specific temperature range, test the experimental circuit within that range to find the best value for R_2. Keep in mind that a change in R_2 value will also change the interbase voltage (V_{B2B1}) which, in turn, will affect the frequency and output voltage.

1-1-3 *Alternate Design Using* UJT *Data Sheets*

Section 1-1-2 describes the procedure for selecting UJT oscillator component trial values "from scratch." It is usually not necessary to design a UJT oscillator "from scratch" if the UJT data sheet is available. Typically, the UJT data sheet will show the UJT connected in a working circuit (or "test" circuit) as a relaxation oscillator. All or part of the values are given, together with the rated output voltage. These values can be used as a starting point for design. For example, if a different frequency is required, use all of the component values shown on the data sheet, except for the value of C_E. Change C_E as necessary to produce the desired oscillation frequency. If a different output voltage is required, change the value of R_1 from that shown on the data sheet. (Use the curves of Fig. 1-9 to find a new trial value for R_1.)

The UJT data sheet will also include absolute maximum ratings (such as power dissipation, temperature range, etc.) and typical electrical characteristics (such as standoff ratio, interbase resistance, etc.). All of these characteristics should be studied carefully before starting design, as is the case with conventional two-junction transistors. For example, if interbase voltage should not exceed 30 V, the source should not exceed 30 V, since the full source voltage could be placed across the bases (if R_1 and R_2 are eliminated, or are very low value).

When the UJT oscillator is to be used as a trigger source for an SCR or other thyristor, the UJT data sheet will often include a special set of curves. The curves show values of R_1 to operate with specific SCRs (by number or type). This type of data makes design of UJT trigger oscillators quite simple. However, the curves generally show only R_1 values for SCRs of one manufacturer (the manufacturer publishing the data sheet). For that reason, use of curves may be somewhat limited.

One data-sheet characteristic often overlooked in design of UJT oscillators is *peak emitter current* or *maximum emitter current* (or whatever similar term is used). When the output voltage across R_1 is connected to a load, most of the load current is drawn through the emitter during the pulse period. To be on the safe side, make sure that the maximum possible load current is at least 10 per cent below the peak emitter current rating of the UJT.

1-1-4 *Synchronization of* UJT *Oscillators*

In many applications, it is necessary to synchronize the UJT oscillator with some external signal (such as a clock or master oscillator in a computer or other digital system). For best results, the UJT oscillator should be designed to operate at a frequency slightly below that of the synchronizing signal. This will ensure that the UJT will not trigger between synchronizing pulses.

A UJT relaxation oscillator can be synchronized by means of either

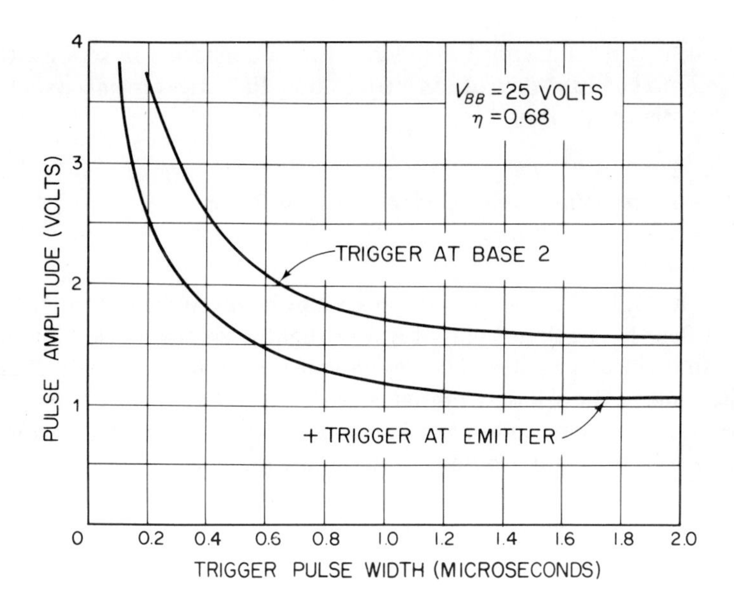

Fig. 1-10 Minimum trigger amplitude as a function of trigger pulse width for turn ON of UJT relaxation oscillator (Courtesy of General Electric Semiconductor Products Department, Syracuse, N.Y.)

positive pulses at the emitter or negative pulses at Base 2. This will raise the emitter voltage, or lower the peak voltage, respectively. Either way, the UJT oscillator will be locked in with the synchronizing signal. In a typical UJT oscillator, a signal of about 1 V is required at the emitter, or about 1.5 V at Base 2 to produce synchronization. These voltages are assumed to have an interbase voltage of 25 V (or less) and a synchronizing signal pulse width of 1 μS or wider. A narrower synchronizing pulse width may require a higher voltage value.

The effect of pulse width on the required trigger amplitude is shown in Fig. 1-10. For pulse widths greater that 1 μS, the required trigger amplitude approaches the d-c conditions. For pulse widths of less than 1 μS, the required pulse amplitude is inversely proportional to the pulse width.

Synchronization of UJT oscillators is discussed more fully in Sec. 1-2 of this chapter.

1-1-5 Protecting UJT *Oscillator Circuits*

If a Base 2 resistor R_2 of the proper value is included in a UJT oscillator circuit, and the maximum data-sheet ratings of peak emitter current and maximum interbase voltage are observed, there should be no damage to the UJT. In general, most UJT circuits assume a peak voltage of 30 V, a C_E of 10 μF or less, and a peak emitter current of 2 A. If any of these factors are exceeded, it is recommended that a resistor be connected in series with C_E. This will limit emitter current. As a trial value, the series resistance should be at least 1 ohm per μF of C_E.

1-1-6 *Alternate Output Connections for Basic* UJT *Oscillator*

As shown in Fig. 1-1, output pulses can be taken from Base 2, as well as from Base 1. The Base 2 output will be slightly less than Base 1 (in amplitude), and will be negative. The design considerations for a Base 2 output are the same as for a Base 1 output.

It is also possible to take an output from the emitter as shown in Fig. 1-1; however, the emitter output is essentially a sawtooth waveform. Use of the basic UJT relaxation oscillator as a sawtooth generator is discussed more fully in Sec. 1-3 of this chapter.

1-2 Unijunction Trigger Circuits for Gated Thyristors

Unijunction transistors provide a simple, convenient means for obtaining a thyristor trigger pulse that is synchronized to the a-c line at a controlled phase angle. This phase control system, using UJTs and thyristors, is one of the most common means of controlling the flow of power to electric motors, lamps, and heaters. With a-c voltage applied to the circuit, the thyristor (SCR, Triac, etc.) remains in its OFF state for the first portion of each half-cycle of the power line voltage. Then, at a time (phase angle) determined by the control circuit, the thyristor switches ON for the remainder of the half-cycle. By controlling the phase angle at which the thyristor is switched ON, the *relative power* in the load may be controlled.

UJT trigger circuits are all based on the simple relaxation oscillator circuit of Fig. 1-11. Note that this circuit is essentially the same as the basic UJT oscillator described in Sec. 1-1, with two exceptions. First, the source voltage V_s is "controllable" in that it can be turned ON and OFF by switch S_1. Second, the Base 1 output is applied to the gate of a thyristor which, in turn, is in series with a load.

The R_T and C_T in Fig. 1-11 form the timing network that determines the time between the application of voltage to the control circuit (represented by the closing of S_1) and initiation of the pulse to the thyristor.

In the circuit of Fig. 1-11, with V_s pure dc, the UJT oscillator is free-running. R_T and C_T determine the frequency of oscillation. The peak of the output pulse voltage is clipped by the forward conduction voltage of the gate-to-cathode diode in the thyristor.

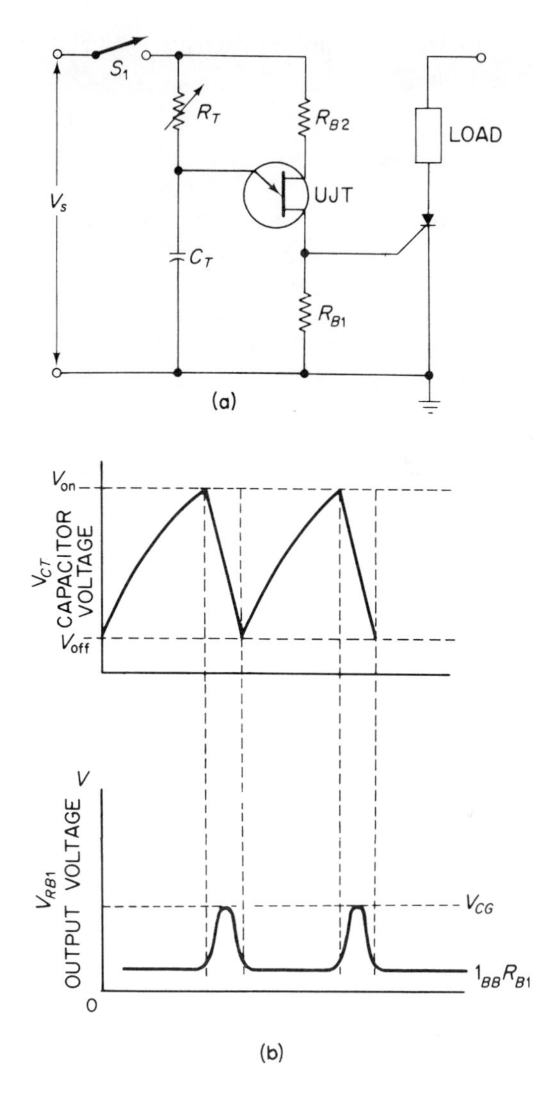

Fig. 1-11 Basic UJT trigger circuit (Courtesy of Motorola Inc., Semiconductor Products Division)

Operation of the circuit may best be understood by referring to the capacitor voltage waveforms shown in Fig. 1-1b, along with the Base 1 output pulse waveform. When power is applied, C_T charges, at the rate determined by its own capacitance and the value of R_T, until capacitor voltage reaches the peak point voltage of the UJT. At that time, the UJT switches into conduction in the normal manner, discharging C_T through R_{B1} and the gate of the thyristor. With V_s pure dc, the cycle repeats immediately. However, in many cases V_s is derived from the anode voltage of the thyristor so that the timing cycle cannot start again until the thyristor is blocking forward voltage, and once again provides V_s.

1-2-1 Design Considerations

During the time in which the capacitor is being charged, current flows through the interbase resistance (r_{BB}) of the UJT. R_{B1} is included in the circuit to provide a path for this current so that the current does not flow through the thyristor gate. This can cause an undesirable turn-on of the thyristor. The value of R_{B1} is selected so that a maximum voltage developed across R_{B1} is less than 0.2 V. For a typical UJT with r_{BB} of 4 to 9 kΩ and a typical operating voltage of 20 V, the value of R_{B1} is

$$R_{B1} = \frac{0.2\, r_{BB(\min)}}{V_s} = \frac{0.2 \times 4\text{K}}{20} = 40 \text{ ohms}$$

As discussed in Sec. 1-1, some UJT data sheets specify a recommended value of Base 1 resistance for a specific type of thyristor. This recommended value should always be used as the first trial value.

Base 2 resistance R_{B2} is necessary only if some degree of temperature compensation is necessary. The value of R_{B2} is found by using the equation or curves discussed in Sec. 1-1. In the case of simple power control circuits,

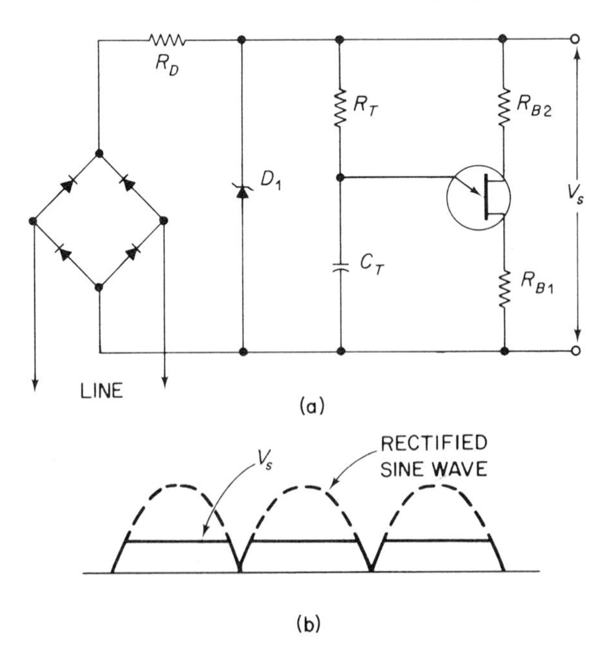

Fig. 1-12 Basic control circuit for synchronizing a UJT trigger (Courtesy of Motorola Inc., Semiconductor Products Division)

particularly in the feedback systems described later in this section, the Base 2 resistance may be left out. In such circuits, the Base 2 of the UJT is connected directly to the positive side of V_s.

It is often necessary to synchronize the timing of the output pulses to power line voltage zero-crossing points. One simple method of accomplishing synchronization is shown in Fig. 1-12. Zener diode D_1 clips the rectified supply voltage, resulting in a V_s as shown in Fig. 1-12b. Since V_{BB} and, thus, the peak point voltage of the UJT drops to zero each time the line voltage crosses zero, C_T is discharged at the end of every half-cycle and begins each half-cycle in the discharged state. Thus, even if the UJT has not triggered during one half-cycle, the capacitor begins the next half-cycle discharged so that the phase angle at which the pulse occurs is directly controlled for each cycle (by the values of C_T and R_T). The Zener diode also provides voltage stabilization for the timing circuit, giving the same pulse phase angle regardless of normal line voltage fluctuations.

1-2-3 Design Example of Basic Half-Wave Trigger Circuit

The most elementary application of the UJT trigger is a half-wave circuit shown in Fig. 1-13. The values of R_T and C_T set the frequency of operation. Note that R_T is made variable so that the phase angle (and thus the power

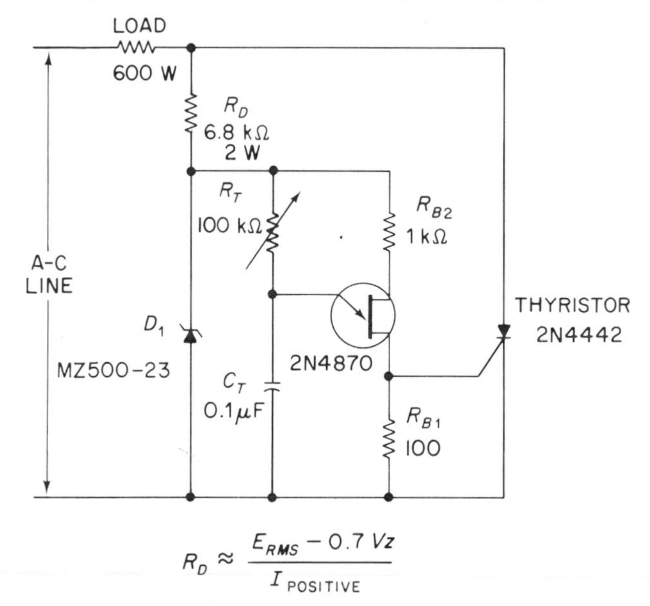

$$R_D \approx \frac{E_{RMS} - 0.7\,Vz}{I_{\text{POSITIVE}}}$$

Fig. 1-13 Half-wave UJT control circuit with typical values for a 600-W resistive load and 60-Hz line (Courtesy of Motorola Inc., Semiconductor Products Division)

applied to the load through the thyristor) can be adjusted. The values of R_T and C_T are suitable for typical 60-Hz line power.

The value of R_{B1} is obtained from the data sheet (for the type of thyristor used), as is the value of R_{B2} (as an optimum value for the anticipated temperature range).

The value of D_1 is determined by the maximum allowable interbase voltage for the UJT. The Zener voltage of D_1 should be slightly less than the maximum allowable UJT interbase voltage. The actual interbase voltage will then be well within the allowable range (due to the additional drop across R_{B1} and R_{B2}).

The value of R_D is selected to limit the current through D_1 so that the diode dissipation capability is not exceeded. Since D_1 is conducting in the regulating mode only during positive half-cycles, the allowable positive current can be calculated when the allowable diode dissipation is divided by one-half the Zener voltage. For example, if D_1 has a 150-mW rating, and the Zener voltage is 24 V, the allowable positive current is 12.5 mA ($24 \times 0.5 = 12$; $0.150/12 = 0.0125$). If the Zener data sheet shows a specific current, usually listed as I_{ZT} or some similar term, use that figure for I_{positive}.

Once the positive half-cycle current is determined, the value of R_D is found by subtracting 0.7 times the Zener voltage from the RMS line voltage and dividing the result by the positive current as shown by the equation in Fig. 1-13.

For example, assume that the Zener voltage is 24 V, line voltage is 115 V, and I_{positive} is 15 mA. R_D is then $115 - (0.7 \times 24)/0.014 = 6.8\ \text{k}\Omega$ (nearest standard value).

Although only the positive half-cycles need be considered in calculating the resistance value, the power rating of R_D must be calculated on the basis of full-wave conduction. (D_1 is conducting on the negative half-cycle acting as a shunt rectifier, as well as providing V_s on the positive half-cycle.)

Using the example of 6.8 kΩ and 15 mA, the minimum power rating of R_D is (I^2R) $15^{-3} \times 15^{-3} = 225 \times 10^{-6}$; $6800 \times 225 \times 10^{-6} = 1.53$ W; use a 2-W value for R_D to provide safe power dissipation.

1-2-4 Providing Half-Wave Control with Switching for Full-Wave Operation

In the circuit of Fig. 1-13, the thyristor is acting both as a power control device and a rectifier, providing variable power to the load during the positive half-cycle, and no power to the load during the negative half-cycle. The circuit of Fig. 1-13 is designed to be a two-terminal control which can be inserted in place of a switch.

If full-wave power is desired at the upper extremes of this control, a switch can be added which will short-circuit the SCR when R_T is turned to its maximum power position. The switch may be placed in parallel with the SCR if the load is resistive. If the load is inductive, the load must be transferred from the SCR to the direct line, as shown in Fig. 1-14.

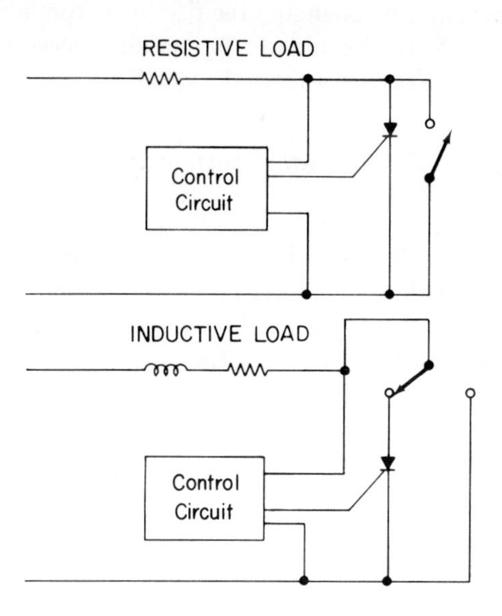

Fig. 1-14 Half-wave controls with switching for full-wave operation (Courtesy of Motorola Inc., Semiconductor Products Division)

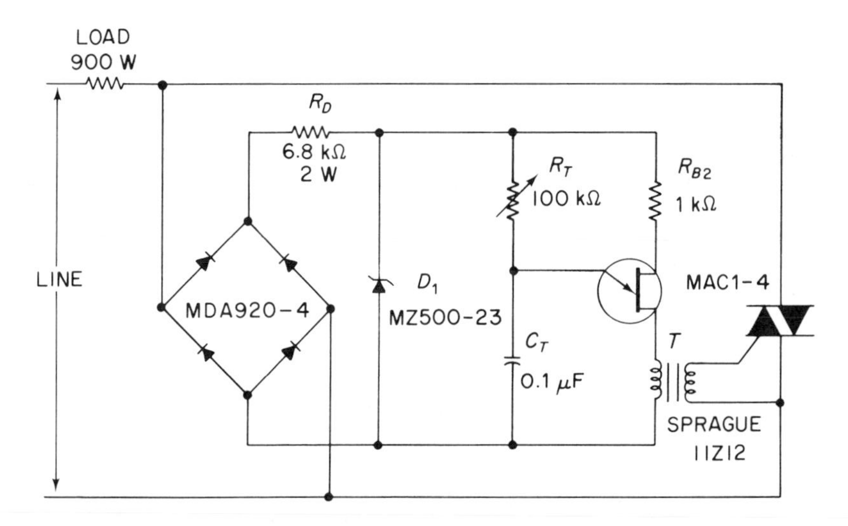

Fig. 1-15 Simple full-wave UJT trigger circuit with typical values for 900-W load and 60-Hz line (Courtesy of Motorola Inc., Semiconductor Products Division)

1-2-5 *Basic Full-Wave Control with* UJT *Trigger*

Full-wave control may be realized by the addition of a bridge rectifier or a pulse transformer, and by changing the thyristor from an SCR to a Triac, as shown in Fig. 1-15. In this circuit, R_{B1} is not necessary since the pulse transformer isolates the thyristor from the steady-state UJT current.

1-2-6 *Full-Wave Control with Constant* UJT *Trigger Output*

Occasionally a circuit is required that will provide constant output voltage regardless of line voltage changes. This can be accomplished by adding a potentiometer in series with R_D, as shown in Fig. 1-16. The value of this potentiometer P_1 should be 10 per cent (or less) of the R_D value.

In operation, P_1 is adjusted to provide reasonably constant output over the desired range of line voltage. As line voltage increases, so does the voltage on the wiper of P_1, increasing interbase voltage (and the peak voltage of the UJT). The increased peak voltage point results in C_T charging to a higher voltage and thus taking more time to trigger. The additional delay reduces the thyristor conduction angle and maintains the average voltage at a reasonably constant value.

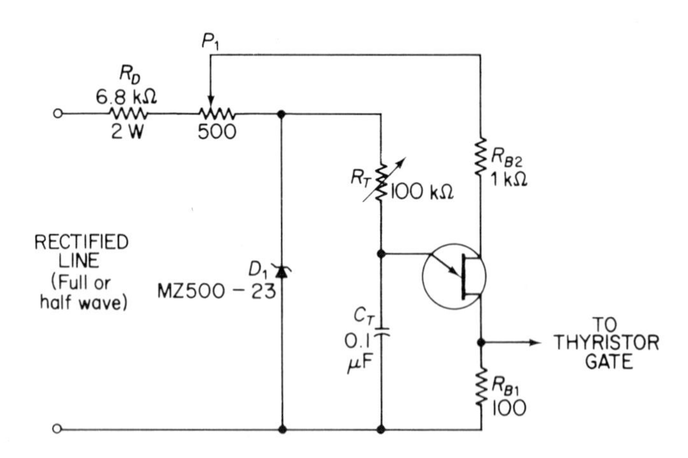

Fig. 1-16 UJT trigger circuit with line voltage compensation (Courtesy of Motorola Inc., Semiconductor Products Division)

1-2-7 UJT *Trigger Circuits with Feedback*

The circuits described thus far have been manual controls; that is, the power output is controlled by a potentiometer (R_T) turned by hand. Simple feedback circuits may be constructed by replacing R_T with heat- or light-

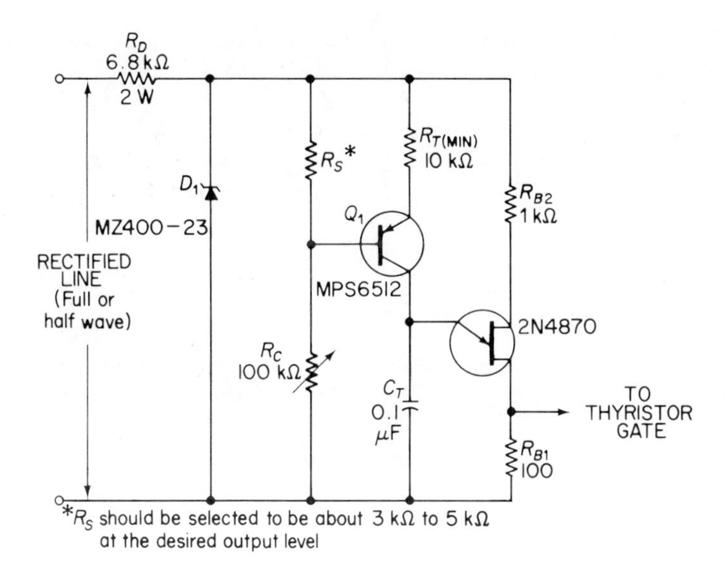

Fig. 1-17 UJT trigger with feedback control (Courtesy of Motorola Inc., Semiconductor Products Division)

dependent sensing resistors; however, such circuits have no means of adjusting the operating levels. By adding a transistor to the basic UJT trigger circuit, it is possible to provide both manual and automatic (feedback) control.

Figure 1-17 shows a feedback control using a sensing resistor R_S for feedback. The sensing resistor may respond to any one of many stimuli such as heat, light, pressure, moisture, or magnetic field. R_S is the sensing resistor and R_C is the manual control resistor that establishes the desired operating point. Transistor Q_1 is connected as an emitter follower such that an increase in the resistance of R_S decreases the voltage on the base of Q_1, causing more current to flow.

Current through Q_1 causes voltage to charge C_T, triggering the UJT at some phase angle. As R_S becomes larger, more current flows into the capacitor; the voltage builds up faster, causing the UJT to trigger at a smaller phase angle; and more power is applied to the load. When R_S decreases, less power is applied to the load. Thus, this circuit is for a sensing resistor that decreases in response to excessive power in the load. If the sensing resistor increases with load power, then R_S and R_C should be interchanged.

1-3 Unijunction Sawtooth Oscillators

The voltage waveform at the emitter of the UJT in the basic relaxation oscillator is a fair approximation of a sawtooth waveform. However, if the emitter output of a UJT oscillator is connected directly to a load (either in-

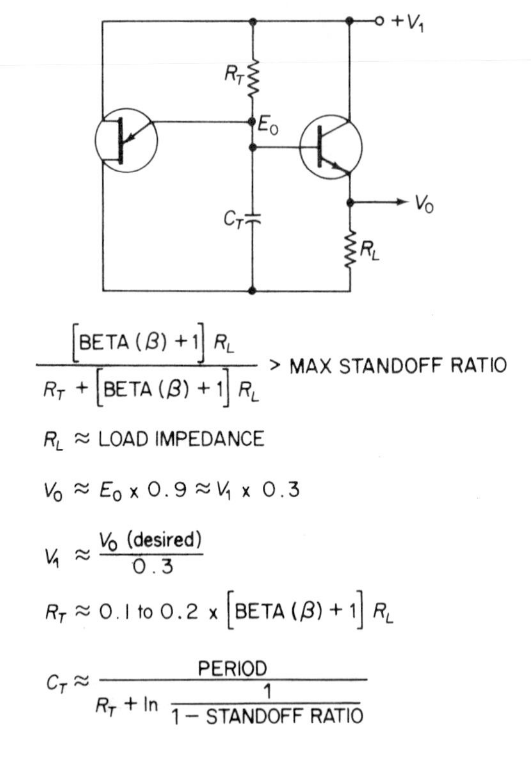

Fig. 1-18 Basic UJT sawtooth oscillator (Courtesy of Motorola Inc., Semiconductor Products Division)

ductive or resistive), the circuit may fail to oscillate. Even if oscillation continues, the waveform will probably be distorted.

The most practical method to couple the emitter output of a UJT oscillator to a load is by means of a direct-coupled emitter follower. Such a circuit is shown in Fig. 1-18. Note that simple direct coupling is made possible by the fact that typical minimum voltage at a UJT emitter $V_{E(MIN)}$ is about 1.2 V. If $V_{E(min)}$ is less than the normal base-to-emitter drop of the transistor, then the waveform across the load R_L will be clipped. However, a typical base-emitter drop of a silicon transistor is about 0.7 V (or considerably less than 1.2 V).

1-3-1 Design Considerations

The primary consideration for the circuit of Fig. 1-18 is the loading effect of the emitter follower. A small amount of loading will shift oscillator frequency. A large amount of loading will stop oscillation. This can be shown by the equivalent circuit of Fig. 1-19.

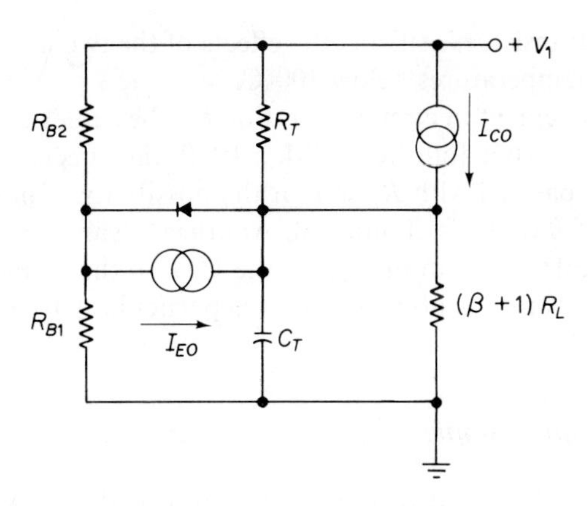

Fig. 1-19 Equivalent circuit of UJT sawtooth oscillator

The loading effect of the emitter-follower stage is approximated by an equivalent circuit $(\beta + 1)R_L$ across capacitor C_T, where β is the d-c common-emitter current gain of the transistor. It is seen from this equivalent circuit that loading will change the frequency of oscillation, since the capacitor charging circuit will be changed by the presence of the resistor $(\beta + 1)R_L$.

To minimize the effects of loading on the frequency, the values of R_L and β should be as large as possible. If the value of β or R_L is too small, the circuit will not oscillate. To ensure oscillation, β and R_L must satisfy the condition:

$$\frac{(\beta + 1)R_L}{R_T + (\beta + 1)R_L} \gg \text{maximum standoff ratio}$$

Often, the value of R_L must be selected to match the impedance of a particular load.

The effects of temperature must also be considered. Two important temperature effects are involved in the use of the emitter-follower stage. First, the variation in β with temperature will change the loading and thus affect the frequency of oscillation. To minimize this temperature effect, $(\beta + 1)R_L$ should be much greater than the value of R_T. The values of C_T and R_T are chosen to provide the desired sawtooth frequency, as described in Sec. 1-1. However, design should start with R_T, by making R_T a value between 0.1 and 0.2 times $(\beta + 1)R_L$. If this results in an impractical value for C_T, increase the value of R_T only as necessary to reduce the size of C_T to practical limits.

The second temperature effect results from the collector leakage current of the junction transistor, shown as I_{CO} in Fig. 1-19. Note that this current adds to the emitter leakage current, I_{EO}, of the UJT. Both leakage currents tend to increase the frequency with increasing temperature. The effect of leakage currents on frequency can be minimized by using a large value for

C_T. If the NPN transistor is silicon, the effects of the two leakage currents can be neglected at temperatures below 100°C.

Some improvement in circuit operation can be achieved by using a PNP transistor as the emitter follower. With a PNP, the effective load resistance, $(\beta + 1)R_L$, is in parallel with R_T so that the possibility of nonoscillation due to a low value of β or R_L is eliminated. Another advantage is that the I_{CO} of the transistor subtracts from the I_{EO} of the UJT so that some degree of temperature compensation is obtained. This is particularly true if a silicon PNP transistor is used.

1-3-2 Design Example

Assume that the circuit of Fig. 1-18 is to provide a sawtooth output of approximately 5 V minimum into a 1000-ohm load. The available power source is 20 V, and a transistor with a β of 50 is to be used. The UJT has a maximum standoff ratio of 0.7.

The value of R_L is 1000 ohms to match the 1 kΩ load.

With R_L at 1 kΩ, and a β of 50, the value of $(\beta + 1)R_L$ is 51 kΩ.

With $(\beta + 1)R_L$ at 51 kΩ, the value of R_T should be between 5.1 kΩ and 10.2 kΩ. Assume a value of 5.1 kΩ.

Substituting these values for comparison against the maximum standoff ratio, we have:

$$\frac{(50 + 1) \times 1000}{5100 + [(50 + 1) \times 1000]} = \frac{51{,}000}{56{,}100} = 0.99$$

Since 0.99 is greater than the maximum standoff ratio of 0.7, oscillation should be sustained with no difficulty.

With an available source voltage (V_1) of 20 V, the V_o output voltage should be approximately 6 V (20 × 0.3) which is greater than the required 5-V output. With 6 V across an R_L of 1000 ohms, there is 6 mA of current through the emitter follower. Of course, the emitter follower must be capable of dissipating this power plus whatever current is passing through the load.

With R_T at 5.1 kΩ select the value of C_T to produce the desired frequency (or period) of operation (as described in Sec. 1-1).

1-3-3 Improving Linearity of UJT *Sawtooth Oscillators*

For many applications, the sawtooth linearity obtained with the basic UJT relaxation oscillator is inadequate. Those UJTs with the *lowest* standoff ratio produce the most linear sawtooth waveforms. However, 10 per cent is about the best linearity that can be obtained with low standoff ratio UJTs.

A number of simple circuit techniques can be used to improve linearity of the sawtooth waveform from a UJT oscillator. Some practical circuits are shown in Fig. 1-20.

Figure 1-20a shows the direct approach of using a higher supply voltage for charging the timing capacitor. This is an inexpensive method of improving linearity if the high voltage supply is available. The circuit of Fig. 1-20a has some disadvantage in that frequency is not as stable as with a single power supply.

Figure 1-20b shows the use of a charging choke to maintain a constant charging current to the timing capacitor. This circuit requires a time constant for the charging circuit which is much greater than the period of oscillation, as shown by the equation. This condition usually produces an impractical size of L at oscillator frequencies of less than 1 kHz.

Figure 1-20c shows the use of the high output impedance of a common-base transistor to maintain a constant charging current for the capacitor. The values of R_T and C_T remain the same as for the basic circuit.

Two variations of a *bootstrap charging* circuit are shown in Figs. 1-20d and 1-20e.

In Fig. 1-20d, a constant voltage is maintained across R_3 by Zener diode D_1 and the emitter follower transistor amplifier stage so that the capacitor charging current is constant over the complete cycle. This circuit is quite economical because it makes double use of the transistor, both as a driver for the bootstrap circuit and as an output amplifier stage. Note that R_4 is returned to a negative voltage. If R_4 is grounded, the current flowing through D_1 would cause clipping at the bottom of the sawtooth waveform.

The values shown for C_1, R_1, R_2, and R_3 provide a sawtooth output at about 2 kHz. These values are selected on the same basis as for the basic circuit. The value of D_1 is 6 V, which is typical for a supply voltage in the 20- to 25-V range. The value of R_4 is chosen to match a given load impedance; however, a change in R_4 value will change the output voltage, all other factors remaining equal. Note that the frequency of the Fig. 1-20d circuit is somewhat dependent upon the supply voltage.

In Fig. 1-20e, the circuit uses a capacitor C_2 in place of the Zener diode. This variation permits the negative supply to be eliminated, thus making frequency largely independent of the supply voltage.

In each of the circuits shown in Figs. 1-20a through 1-20e, the linearity is limited by the loading effect of the output stage so that it will not be possible to increase linearity beyond a certain value. This value is set by $(\beta + 1)R_L$. The circuit of Fig. 1-20f shows a method of compensating for both the loading of the output stage and the variable charging current of the timing capacitor. Resistor R_3 and capacitor C_2 act as an integrating network of the waveform. By varying the values of R_3 and C_2, the output waveform can be made concave upward, concave downward, or linear. In a practical application, the circuit is assembled in breadboard form, using a potentiometer for

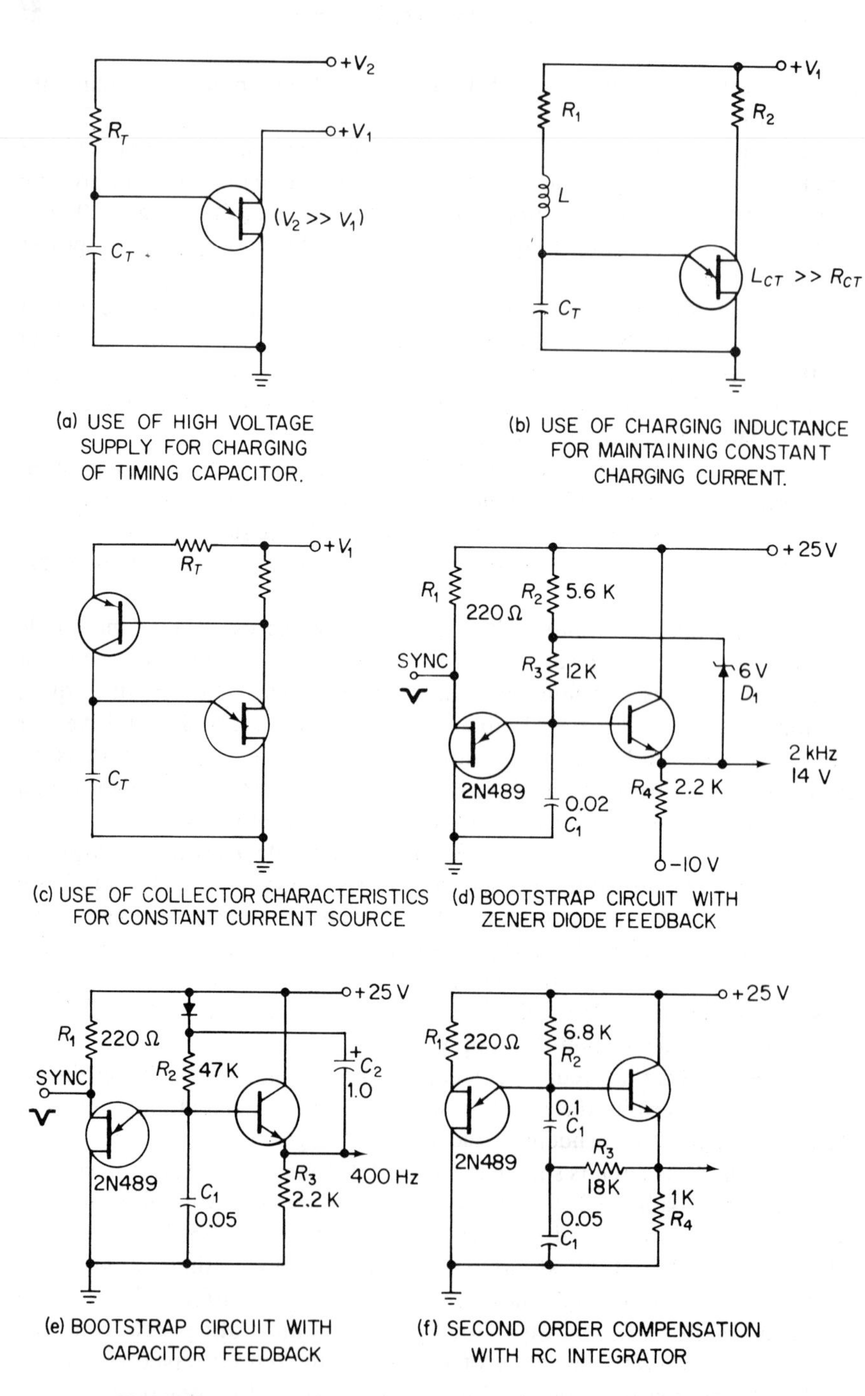

Fig. 1-20 Circuits for improving linearity of UJT sawtooth oscillators (Courtesy of General Electric Semiconductor Products Department, Syracuse, N.Y.)

R_3. The output is monitored on an oscilloscope, and the value of R_3 is adjusted to provide the most linear waveform.

1-4 Unijunction Bistable Circuits

The negative resistance characteristics of a UJT make it possible to use a single UJT as the active element of a bistable switching device. The design and analysis of UJT bistable circuits are quite different from those of conventional two-junction transistors. For that reason, we shall discuss the basic UJT bistable circuit before going into specific design considerations and design example.

The basic form of UJT bistable circuit is shown in Fig. 1-21, together with the corresponding emitter characteristic curve and load line. The two stable operating points of this circuit are indicated by points A and B on the emitter characteristic curve where the load line formed by R_L and V_1 inter-

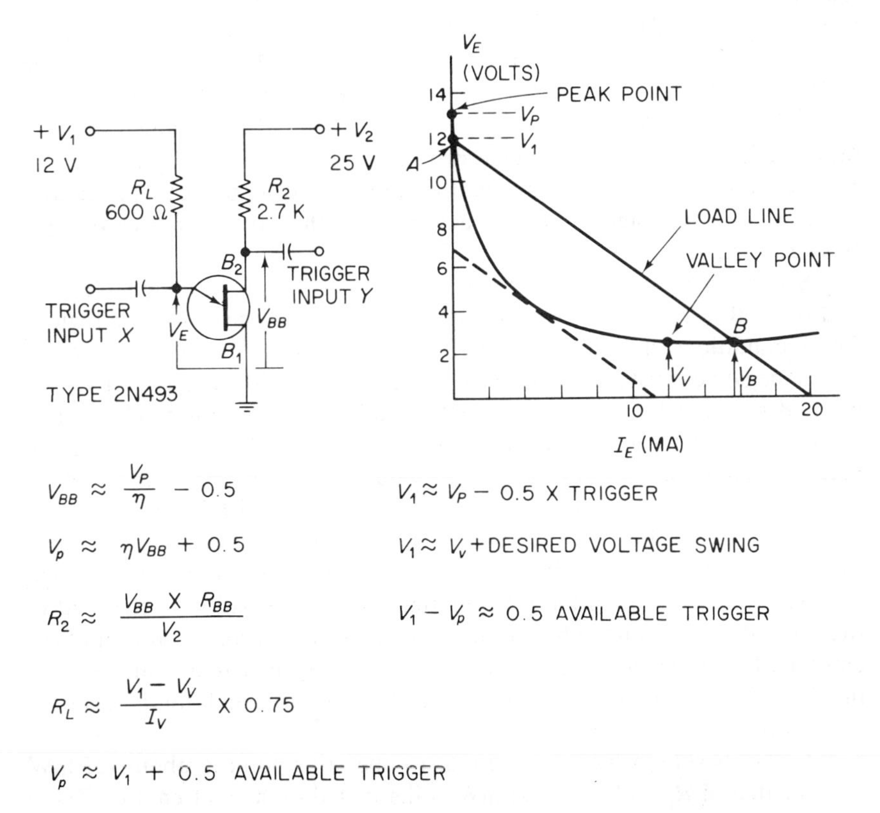

Fig. 1-21 Basic UJT bistable circuit (Courtesy of General Electric Semiconductor Products Department, Syracuse, N.Y.)

sects the curve in the regions of positive slope. Point A is in the cutoff region where the emitter diode is reverse-biased. Point B is in the saturation region, where the emitter is conducting and the emitter voltage is low. Note that neither state is in the negative-resistance region, as is required for a UJT oscillator described in Sec. 1-1. The slope of the emitter characteristic curve in the saturation region is generally less than 40 ohms. In contrast, the slope in the cutoff region is generally greater than 10 megohms.

1-4-1 Design Considerations

Two conditions must be met for reliable operation of a bistable UJT circuit. First, the supply voltage V_1 must be less than the peak-point voltage V_p, by a margin determined by the operating temperature range and other design considerations. If this condition is not met, point A will cease to be a stable state for the circuit whenever V_p is less than V_1. As shown by the equations, the amplitude of the available trigger should also be considered when selecting a value for V_1. Unless there is some other requirement to the contrary, use one-half the trigger value for the differential between V_p and V_1. For example, if V_p is 15 V and the available trigger is 2 V, then V_1 should be 14 V (2 V $\times$ 0.5 = 1 V; 15 V − 1 V = 14 V). Keep in mind that V_p is not a fixed value but is determined by such factors as interbase voltage, standoff ratio, and emitter forward voltage drop ($V_p = \eta V_{BB} + V_D$). V_p can be calculated from the data-sheet information; however, in a practical application, the UJT must be tested to find the actual V_p (with a specific source voltage, Base 2 resistance, etc.).

Second, the resistor R_L must have a value which is small enough to ensure that point B will fall to the right of the valley point (in the saturation region) for all conditions of operation. If point B falls to the left of the valley point (in the negative-resistance region), then the circuit could be regenerative and point B may not be an unconditionally stable operating point. This is just the opposite of the requirement for a relaxation oscillator as described in Sec. 1-1. Thus, the absolute minimum value of R_L for a UJT oscillator is slightly larger than the absolute maximum for a UJT bistable circuit.

The circuit of Fig. 1-21 can be considered as being in the OFF state when the operating point is at A since there is only a small current flowing through the load R_L and through the UJT emitter. Power dissipation in the load in the OFF state is determined by the size of R_L and by the leakage current of the emitter which is generally less than 1 μA. Thus, for the circuit values shown in Fig. 1-21, power dissipation in the load in the OFF condition is less than 10^{-9} watt.

In the ON state, which corresponds to operating point B, about 135 mW is dissipated in R_L and about 45 mW is dissipated in the UJT emitter. These figures can be found by noting the voltage and current at point B (approximately 3 V and 15 mA, or 45 mW) to find UJT emitter dissipation. The same 15 mA flows through R_L, producing an approximate 9-V drop (from V_1 of 12 V to about 3 V at point B). Thus, dissipation in R_L is 9 V $\times$ 15 mA = 135 mW, or $0.015^2 \times 600 = 135$ mW.

The values shown in Fig. 1-21 are by no means the upper limit of the power that can be switched by a UJT. With the proper choice of supply voltages and load resistances, it is possible to switch up to about 1.5 W with a typical UJT.

To switch the bistable UJT ON, a trigger may be applied at input X or input Y, or both. If the trigger is applied at point X, the trigger must be positive with an amplitude greater than $(V_p - V_1)$. To ensure turn-on, the trigger should be approximately twice the differential between V_p and V_1. Triggering at input X essentially *raises* the load line on the characteristic curve.

If the trigger is applied at input Y, the trigger must be negative and an amplitude greater than $(V_p - V_1)/\eta$. Triggering at input Y *lowers* the emitter characteristic curve by changing the effective value of interbase voltage V_{BB}.

To switch the bistable UJT OFF, it is necessary to apply a negative trigger pulse at input X. The required amplitude is equal to $V_B - V_v$, which is typically 0.1 V, and rarely over 0.5 V. Note that the emitter input impedance of the UJT in the ON state is quite low. Thus, the output impedance of the trigger source must also be low.

Although it is possible to turn the bistable circuit ON by means of a trigger at Base 2 (input Y), it is not possible (without extensive modification) to turn the circuit OFF with a trigger at Y. To provide turn-off with a trigger at Y, the values must be chosen so that point B is slightly to the left of the valley point, just inside the negative-resistance region. If a positive pulse is applied at input Y under these conditions, the circuit will be unstable and the UJT will turn OFF. Such an arrangement is generally not practical, so point Y is rarely used for turn-off.

The bistable UJT may also be turned OFF by momentarily reducing the voltage V_1. This will, in effect, cause the position of the load line on the characteristic curve to shift. The requirement for unconditional turn-off is met when the load line intersects the emitter characteristic curve only in the cutoff region. This condition is shown by the dotted line in Fig. 1-21, and indicates that V_1 must be reduced to less than 7 V to ensure turn-off.

If the load resistor R_L is connected between Base 1 and ground, a variation of the basic UJT bistable circuit is obtained. This variation has certain advantages in some applications. The circuit may be turned ON by a negative pulse applied at Base 1 or at Base 2. The circuit can also be turned OFF by a positive pulse at Base 1. The current gain between emitter and Base 1 permits higher switching efficiency in this variation. However, the variation has some disadvantage in that current flows through the load in the OFF condition through the interbase resistance (typically between 4 kΩ and 9 kΩ).

In addition to specific voltage requirements, the current capabilities of the trigger source must be considered. Essentially, emitter current must be greater than the peak point emitter current I_p if the UJT is to turn ON. The peak point current is typically low (nominally about 4 μA with a V_{BB} of 25 V at a temperature of 25°C). In extreme cases (low temperatures of −60°C and an interbase voltage of 4 V), the peak point current of a typical UJT might rise to 15–20 μA.

If *trigger pulses are used* instead of a d-c trigger, the required pulse amplitude increases as the pulse width decreases (a condition typical for most pulse circuits). The graph of Fig. 1-10 (used for UJT oscillator triggers) can also be applied to UJT bistable circuits. Note that trigger pulse widths as low as 1 μS may be used without appreciable increase in required amplitude.

Trigger requirements for turning off a UJT bistable circuit are quite complex. A simple rule is that *turn-off can be made faster when the value of the emitter current immediately before turn-off is decreased.* Also, faster turn-off is obtained if the emitter is driven negative with respect to Base 1, and if Base 2 is kept out of saturation (as is discussed later in this section). Current may flow either into or out of the emitter on turn-off. If the emitter is driven negative with respect to Base 1 on turn-off, the emitter diode exhibits storage effects similar to those of some types of junction diodes.

Because of these complex turn-off characteristics, the measurement of UJT turn-off is difficult. Before the UJT is turned OFF, the conductivity of the region between the emitter and Base 1 is high because of excess concentration of holes and electrons. On turn-off, this charge concentration decreases to its low initial value, resulting in corresponding increase in peak-point emitter voltage V_p. This recovery effect is difficult to measure since it cannot be observed directly at the terminals of the UJT.

A circuit for indirect measurement of UJT recovery characteristics is shown in Fig. 1-22. Two pulse generators are used to produce the waveforms

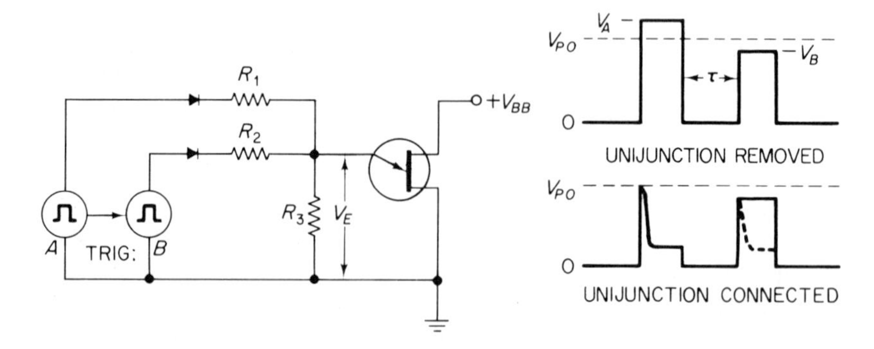

Fig. 1-22 Test circuit for measurement of peak-point voltage recovery time with typical emitter voltage waveform (Courtesy of General Electric Semiconductor Products Department, Syracuse, N.Y.)

shown. Generator A produces a pulse at the emitter of the UJT. The amplitude of the emitter pulse V_A is greater than the peak-point voltage V_{po}. The emitter pulse acts as a trigger and turns the UJT ON for the duration of the pulse. At a time interval T, after the end of the first pulse, a second pulse with an amplitude V_B is produced at the emitter by generator B. This second pulse causes the UJT to fire (dotted line), provided that the instantaneous value of V_p less than V_B. By varying the time delay (T) and the pulse amplitude (V_B), it is possible to measure the recovery characteristic of the peak-point voltage.

A plot of the recovery characteristics is shown in Fig. 1-23. Two different values of initial emitter current are shown. It has been found that, from unit to unit, the recovery time varies inversely with emitter saturation voltage. This condition indicates that recovery time is determined chiefly by the effective carrier lifetime.

Recovery characteristics of the peak-point voltage are of considerable value from the standpoint of practical circuit design. The time constant of the circuit used for turn-off must be designed so that emitter voltage rises

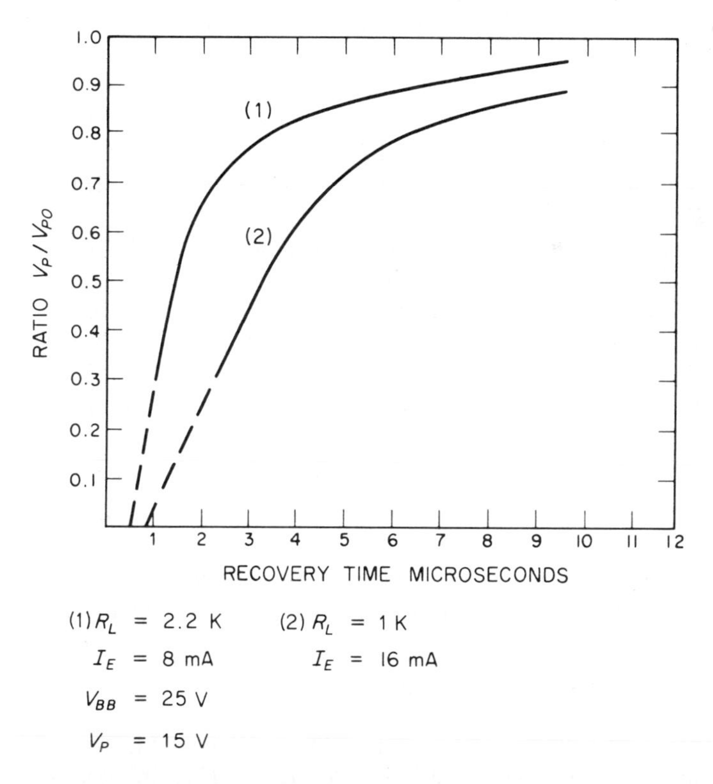

Fig. 1-23 Recovery characteristics of peak-point emitter voltage for two values of emitter current (Courtesy of General Electric Semiconductor Products Department, Syracuse, N.Y.)

more slowly than V_p. If not, the emitter will conduct before recovery is complete, and the UJT will not turn off.

The value of R_2 in Fig. 1-21 sets the level of interbase voltage (and thus the level of V_p, since V_p is determined by interbase voltage, standoff ratio, and emitter forward voltage drop). An increase in R_2 decreases interbase voltage (and V_p). It is possible to set the level of V_p, and thus set the voltage swing of the emitter, by selection of the R_2 value.

1-4-2 Design Example

Assume that the circuit of Fig. 1-21 is to be used as a basic UJT bistable element. The available supply voltage V_2 is 25 V. Thus, the available V_1 voltage is anything up to 25 V. The UJT has an interbase resistance of 4.9K, a typical valley point of 3 V, an emitter drop V_D of 0.5 V, a standoff ratio of 0.9, and is capable of dissipating 300 mW without heat sinks. The desired output voltage swing is 9 V, with an available trigger of 2 V.

With a V_v of 3 V, and a desired voltage swing of 9 V, V_1 is 12 V. With V_1 at 12 V, and the available trigger at 2 V, V_p is 13 V (2 V $\times$ 0.5 = 1 V; 12 V + 1 V = 13 V).

With V_p at 13 V, the standoff ratio of 0.9, and a V_D of 0.5 V, V_{BB} is 14 V (13/0.9 = 14.4 V; 14.4 V − 0.5 = 13.9 V; rounded off to 14 V).

With V_{BB} at 14 V, V_2 at 25 V, and R_{BB} (interbase resistance) at 4.9 kΩ, R_2 is approximately 2.7 kΩ (14 $\times$ 4900 = 68,600; 68,600/25 = 2744, rounded to 2.7 kΩ).

There are two simple methods for finding the value of R_L. If a curve such as shown in Fig. 1-21 is available, draw a load line from V_1 to point B. Note where the load line intersects the emitter current axis (I_E). In the example of Fig. 1-21, the line intersects I_E at 20 mA. Thus, the value of R_L is 600 ohms (12/0.020 = 600).

If the curve is not available or it is impractical to draw a load line, use the equation of Fig. 1-21 for R_L. For example, assume a value of 11 mA for I_v. The value of R_L is approximately 600 ohms, found as follows:

$$\frac{V_1 - V_v}{I_v} \times 0.75 = \frac{12 - 3}{0.011} \times 0.75 = 818 \times 0.75 \approx 600$$

With a 3-V drop across the UJT, and an approximate 15-mA current, UJT dissipation is approximately 45 mW (well below the 300 mW maximum).

1-4-3 Alternate UJT *Bistable Circuits*

In Fig. 1-24, an alternate version of the bistable circuit illustrates some important factors involved in UJT bistable circuits. This circuit uses a clamping diode to hold the emitter voltage below the peak-point voltage. The UJT is turned ON by a *negative* trigger at Base-2. When the UJT is ON, the clamping diode is back-biased, and resistance R_1 serves as the emitter load. Note that in the ON state, the UJT is biased in the negative-resistance region (point B). This will be a stable state if the capacitance (dotted lines) between the emitter and Base 1 is kept below a certain critical value. The critical value of capacitance depends upon the bias point B and increases as the valley point is approached. As a rule, the value of the critical capacitance will generally be greater than 50 pF if the bias point B is below 5 V.

To find the value of R_1 for the circuit of Fig. 1-24, draw the load line from the 25-V point, through point B, to the I_E axis line. In the example of Fig. 1-24, the load line intersects I_E at about 7.5 mA. If drawn from the 25-V point of V_E, the approximate value of R_1 is 3.3 kΩ ($25/0.0075 \approx 3.3$ kΩ).

If the UJT is biased in the negative-resistance region in the ON state, the power required for turn-off, and the turn-off time, will both be greatly reduced. A technique for turn-off is shown in the circuit of Fig. 1-25. If V_3 is

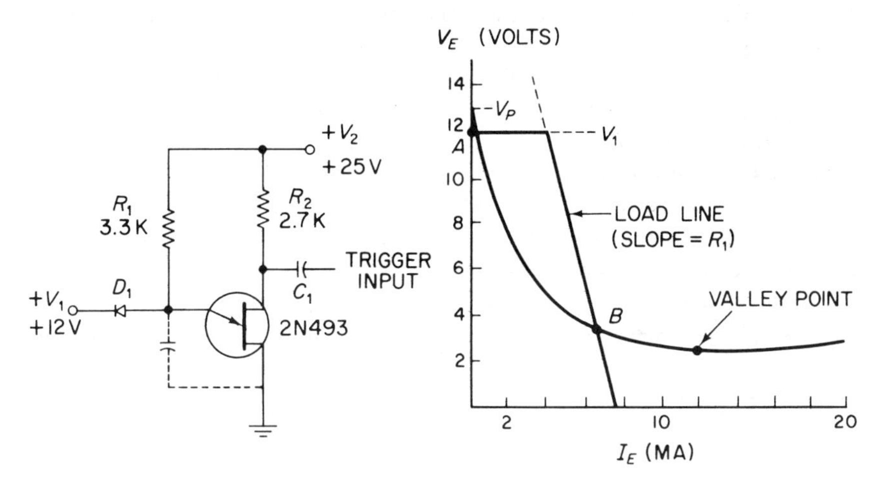

Fig. 1-24 UJT bistable circuit with emitter voltage clamping (Courtesy of General Electric Semiconductor Products Department, Syracuse, N.Y.)

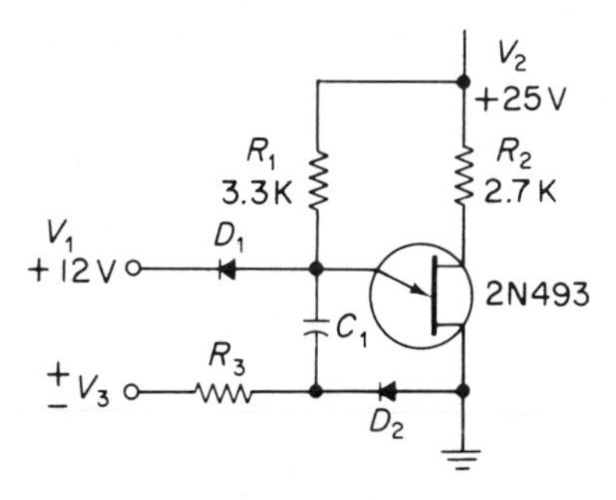

Fig. 1-25 UJT bistable circuit with bias (Courtesy of General Electric Semiconductor Products Department, Syracuse, N.Y.)

negative, the diode D_2 is reverse-biased and the dynamic resistance in series with capacitor C_1 is high so that the UJT is stable at the bias point B. If V_3 is made positive, the diode D_2 is biased in the forward direction and its dynamic resistance will decrease as the forward current increases. When the total dynamic resistance around the loop D_2-C_1-emitter-Base-2 becomes negative, the circuit is regenerative and the UJT turns OFF.

Using this type of circuit has several advantages. Turn-off of the UJT is accomplished with a positive signal rather than the negative signal that is required in other circuit configurations. Because of this, both turn-on and turn-off may be obtained by pulses of the *same polarity and amplitude at the same input*. Thus, a train of identical pulses will turn the UJT ON and OFF.

For turn-on, it is, of course, necessary to have some resistance in series with the diode D_1 and the supply V_1. Turn-off power requirements are quite low. As bias point B of the ON state (Fig 1-24) approaches the peak point, the turn-off current approaches the peak-point current (typically a few microamperes).

Another version of the UJT bistable circuit is shown in Fig. 1-26. Here the slope of the load line between points A and C is determined by the parallel combinations of R_1, R_3, and R_4. When the UJT is on, the voltage at Z is equal to V_C. The diode is thus back-biased by a voltage $V_C - V_B$ so the emitter is decoupled from point Z and the capacitor does not cause the UJT to be unstable at the bias point B. Only a small negative trigger is required at input X to turn the UJT OFF.

Turn-on and turn-off may be obtained with pulses of one polarity if a small coupling capacitor C_1 is used (typically, 0.1 μF or smaller). This results

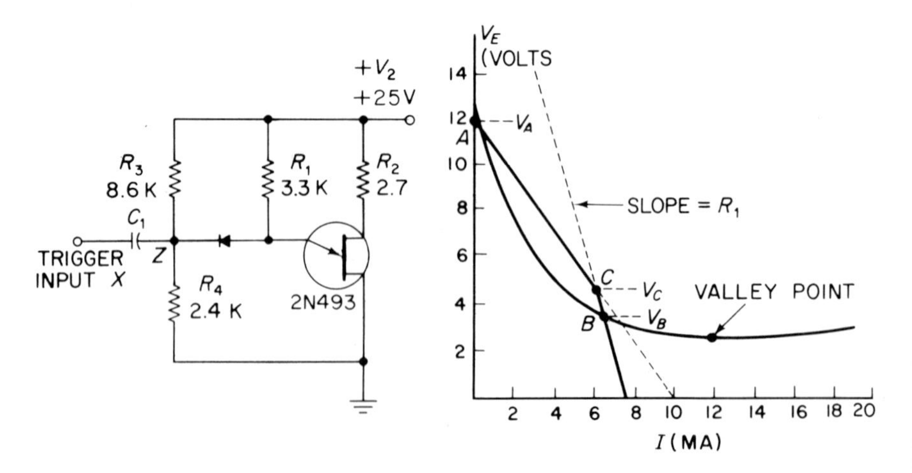

Fig. 1-26 UJT bistable circuit with diode decoupling (Courtesy of General Electric Semiconductor Products Department, Syracuse, N.Y.)

in differentiation of the trigger pulse so that, with a positive trigger pulse, turn-on occurs at the leading edge of the trigger pulse, and turn-off occurs at the trailing edge.

1-5 Unijunction Trigger for Multivibrators

Conventional transistor multivibrators have a number of serious limitations which restrict their use in timing circuits and rectangular wave generators. For example, the cross-coupling capacitors must serve as the timing capacitors and thus are required to have large values if long timing periods or heavy loads are required. Often, the required values of capacitance are so high that electrolytic capacitors must be used, severely limiting the accuracy and stability of the timing periods. The coupling capacitors also result in considerable distortion of the collector voltage waveform so that optimum rectangular waveforms cannot be obtained. This is particularly true for an unsymmetrical multivibrator which has an appreciable difference in the lengths of the two parts of the timing cycle.

These problems can be overcome by combining the UJT oscillator with a conventional two-junction transistor multivibrator circuit. The UJT oscillator provides the timing trigger, while the multivibrator transistors act as the switches. Such a circuit permits independent adjustment of the length of the two parts of the timing cycle, over a range as large as 1000 to 1. The circuit also offers the advantage of ideal rectangular waveform output and good timing stability, and requires only one timing capacitor of minimum size.

1-5-1 Design Considerations

Before going into the combined UJT trigger and multivibrator (often known as a "hybrid multivibrator") circuit, we shall discuss the conventional multivibrator. However, there will be no detailed discussion of basic multivibrators here, since they are described fully in the author's *Handbook of Simplified Solid-State Circuit Design*.

For the purpose of comparison, a two-transistor multivibrator of conventional design is shown in Fig. 1-27. This circuit is of the high-current type. That is, the emitter resistors are about half the collector resistor values, resulting in very high switching currents through the transistors. This requires transistors with a higher current capability (and higher power dissipation), but produces a circuit with high-frequency stability. That is, the circuit will maintain frequency to about 1 part in 10^4, in spite of large changes in supply voltage.

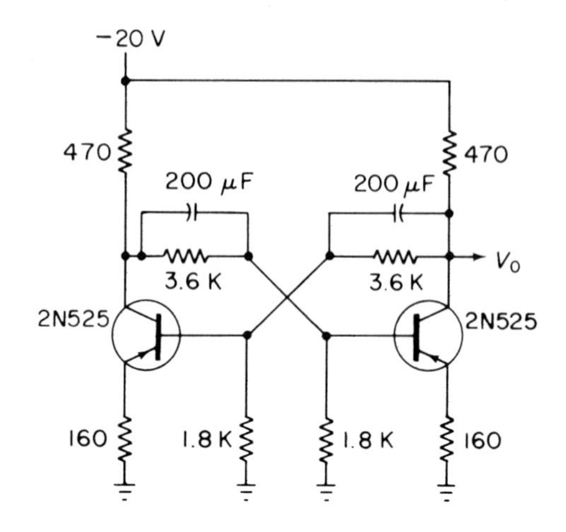

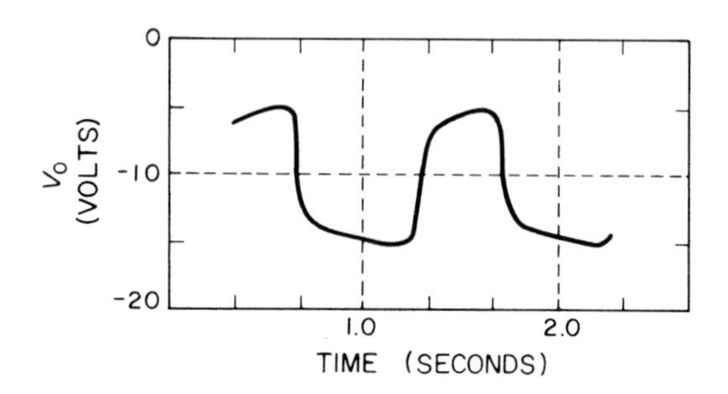

Fig. 1-27 Typical *PNP* multivibrator with collector voltage waveform (Courtesy of General Electric Semiconductor Products Department, Syracuse, N.Y.)

The same circuit could be used as a low-current type where the collector resistors are about 10 times the emitter resistance value. This will require a lower current (and power dissipation) capability for the transistors, but will result in frequency changes with variations of the supply voltage.

With either type of circuit, the output is supposed to be a symmetrical square wave. However, as the waveforms of Fig. 1-27 show, the collector voltage waveform for this circuit deviates considerably from an ideal square wave. Output can be taken from either half of the circuit. Likewise, either half of the circuit can be triggered, if desired. The circuit can be either free-running, where the frequency is determined by the RC time constant, or driven by a trigger souce.

Symmetrical Hybrid Multivibrator

The simplest version of the hybrid multivibrator circuit is the symmetrical multivibrator or square wave generator shown in Fig. 1-28. In this circuit, the two PNP transistors are used in a simple saturating flip-flop designed in the normal manner. The UJT triggers the flip-flop from one state to the other by means of the negative trigger pulses developed across R_A. Collector or base triggering of the flip-flop can also be used, but the emitter triggering method is generally simpler and requires fewer components.

Note that the timing is controlled by the UJT oscillator time constants (R_T and C_T), rather than the multivibrator time constants. The values of the multivibrator circuit are chosen so that the transistors can be driven into saturation by pulses from the UJT. The values shown for the multivibrator should be satisfactory for any period greater than about 100 μS.

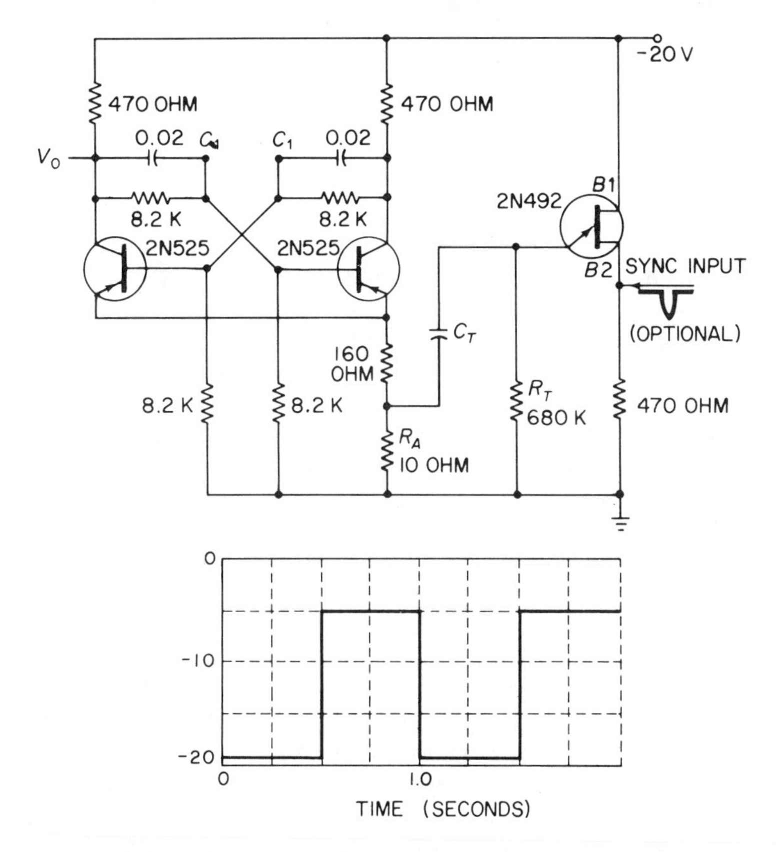

Fig. 1-28 Typical hybrid (UJT-*PNP*) multivibrator with collector voltage waveform (Courtesy of General Electric Semiconductor Products Department, Syracuse, N.Y.)

In comparing the circuit of Fig. 1-28 with that of Fig. 1-27, the improvement in the output voltage is apparent. Also, note that the conventional circuit requires two 200-μF capacitors, whereas the hybrid circuit requires only one 1-μF capacitor (since it is practical to use a very large value of R_T to achieve the 1-second time constant). Such values make it possible to use nonelectrolytic capacitors in the hybrid circuit. Another apparent advantage of the hybrid circuit is that perfect symmetry of the waveform is obtained without requiring the use of a balance control.

If a synchronizing signal is desired, it is best to use negative triggers at Base-2 as shown in Fig. 1-28. Typically, the trigger pulse width should be greater than 0.5 μS.

No design example is given for the symmetrical hybrid multivibrator circuit. The values shown for the multivibrator portion should be satisfactory for any frequency between about 1 Hz and 10 kHz. To obtain a specific frequency of operation within this range, simply change the value of R_T (and C_T, if necessary). (Refer to Sec. 1-1.)

Unsymmetrical Hybrid Multivibrator

If an unsymmetrical waveform is desired, the hybrid circuit may be modified to that shown in Fig. 1-29. This circuit also illustrates the use of an NPN transistor flip-flop. Two separate charging resistors R_{T1} and R_{T2} are used, with each connected to corresponding collectors of the multivibrator through isolating diodes D_1 and D_2. Assume that initially transistor Q_1 is OFF. The voltage at the collector of Q_1 will then be about 12 V, and the voltage at the collector of Q_2 will be about 2 V. The capacitor C_T is charged through R_{T1} until the UJT fires. At this time, the flip-flop will be triggered to the opposite state and Q_2 is OFF. The capacitor C_T is then charged through R_{T2}, and diode D_1 isolates R_{T1} from the timing circuit.

Resistors R_{T1} and R_{T2} determine, independently, the lengths of the two parts of the timing period. If potentiometers are used in place of R_{T1} and R_{T2}, the lengths of the two parts of the timing period will be adjustable over a range as large as 1000 to 1. Note that it is advisable to use fixed resistors (about 3K) in series with the potentiometers to prevent burning out the UJT when the potentiometers are set at their minimum values.

If independent adjustment of the two parts of the period is not required, the circuit can be simplified by replacing R_{T2} with a 130K resistor, eliminating D_1, and connecting R_{T1} from the emitter of Q_3 to the positive supply.

If a multivibrator with a fixed period but a variable duty cycle is desired, R_{T1} and R_{T2} may be replaced by a single potentiometer, with its center tap connected to the UJT emitter, as shown on Fig. 1-29.

In designing unsymmetrical multivibrators with NPN transistors, the collector voltage of the transistors in the ON condition must be less than 2 V,

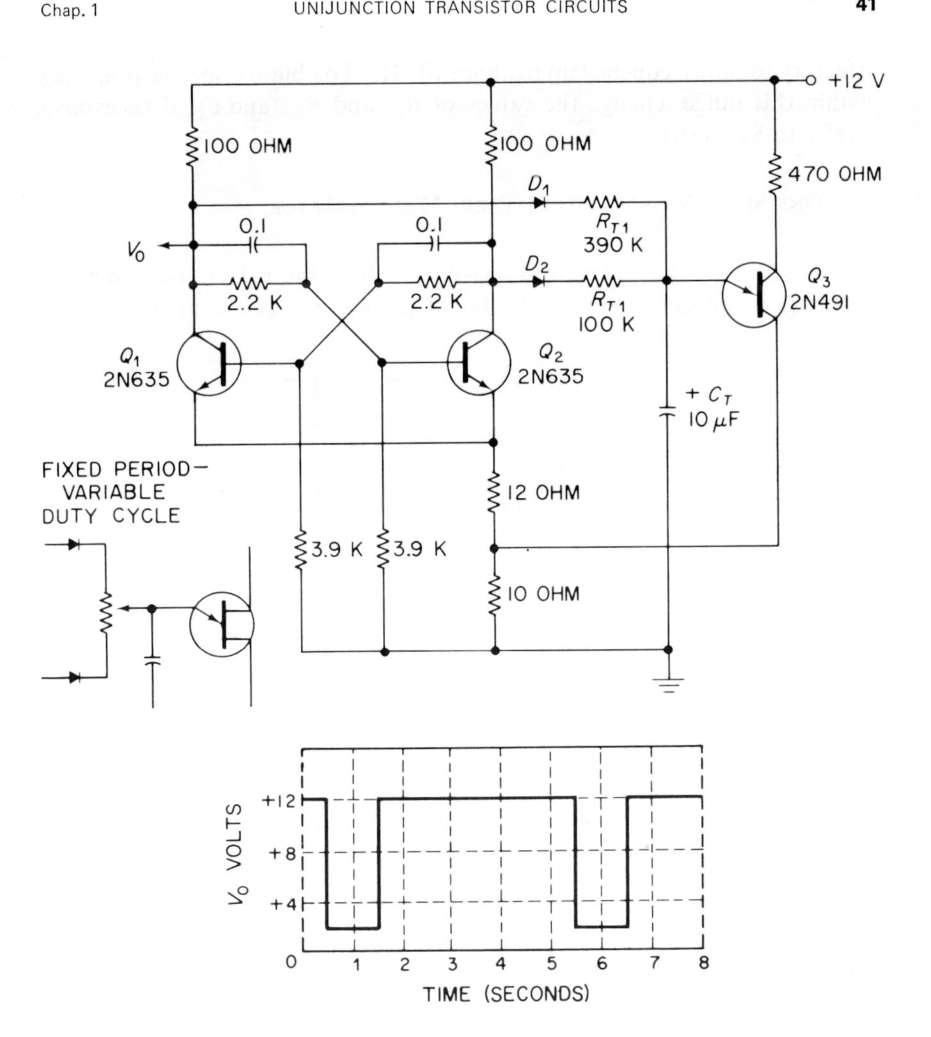

Fig. 1-29 Unsymmetrical hybrid (UJT-*NPN*) multivibrator with collector voltage waveform (Courtesy of General Electric Semiconductor Products Department, Syracuse, N.Y.)

relative to the Base-1 terminal of the UJT. Otherwise, the effect of R_{T1} and R_{T2} on the two parts of the period will not be completely independent. This requirement can be met with the circuit of Fig. 1-29 by interchanging the 10-ohm and 12-ohm resistors, connecting the lower terminal of C_T to their common point, and connecting Base 1 to the emitters of Q_1 and Q_2. Note that this problem does not arise with PNP transistors, or with flip-flop circuits using a separate base bias supply.

Again, circuit timing is controlled by the UJT oscillator time constants. The values shown for the multivibrator portion of the circuit should be satisfactory for any frequency up to about 10 kHz. To obtain a specific frequency within this range, change the values of R_{T1} and R_{T2} (and C_T, if necessary). (Refer to Sec. 1-1.)

One-Shot (Monostable) Hybrid Multivibrator

Figure 1-30 shows a hybrid one-shot multivibrator. Here the timing resistor R_T is connected to the collector of Q_1, and the quiescent state for the

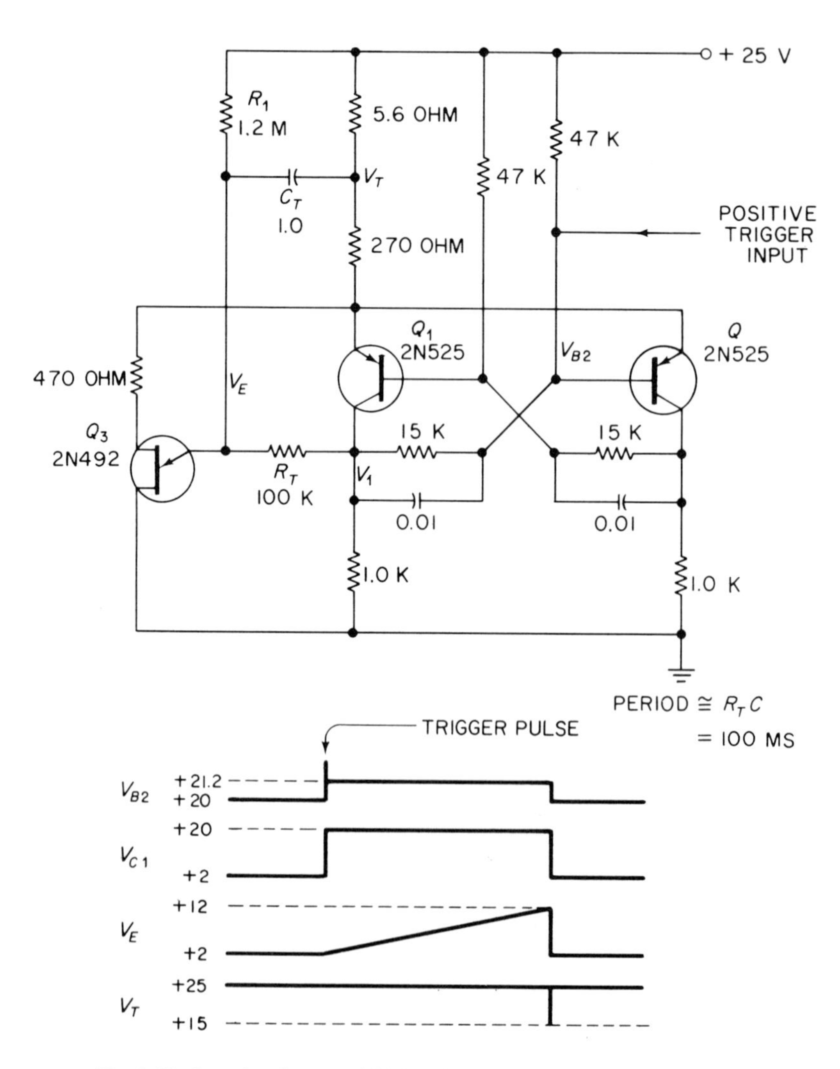

Fig. 1-30 One-shot (monostable) hybrid multivibrator with typical waveforms (Courtesy of General Electric Semiconductor Products Department, Syracuse, N.Y.)

circuit occurs when Q_1 is OFF and Q_2 is ON. A positive trigger applied at the base of Q_2 turns Q_2 OFF and causes the voltage at the collector of Q_1 to increase to about 20 V. Capacitor C_T is then charged through R_T. At the end of the timing cycle, the UJT fires and triggers the flip-flop to its quiescent state.

Note that once the timing cycle is started by a trigger pulse at the base of Q_2, additional trigger pulses at this input will not affect the timing cycle. Another important advantage of the circuit is apparent from the waveforms shown in Fig. 1-30. At the end of the timing cycle, all the voltages in the circuit have been restored to their initial values. For this reason, the timing period is independent of the trigger rate and duty cycle. This permits duty cycles of over 95 per cent, if desired, and is in contrast with conventional one-shots where duty cycles are generally limited to less than 70 per cent (to maintain timing accuracy).

Note that resistor R_1 is used to bias the emitter of the UJT to about 2 V when the circuit is in the quiescent state. This ensures that the emitter voltage returns exactly to its initial value at the end of the timing cycle.

An alternate one-shot hybrid multivibrator is shown in Fig. 1-31. In this circuit, the UJT is normally off. The circuit operation is as follows: When an input trigger pulse of positive polarity is applied to the base of Q_1, Q_1 will turn ON and Q_2 will turn OFF. A voltage is then applied to charge capacitor C_E and, when the voltage on C_E reaches V_p, the UJT Q_3 fires. The positive voltage developed across the 27-ohm resistor is used as the output signal and as a trigger signal to the base of Q_2.

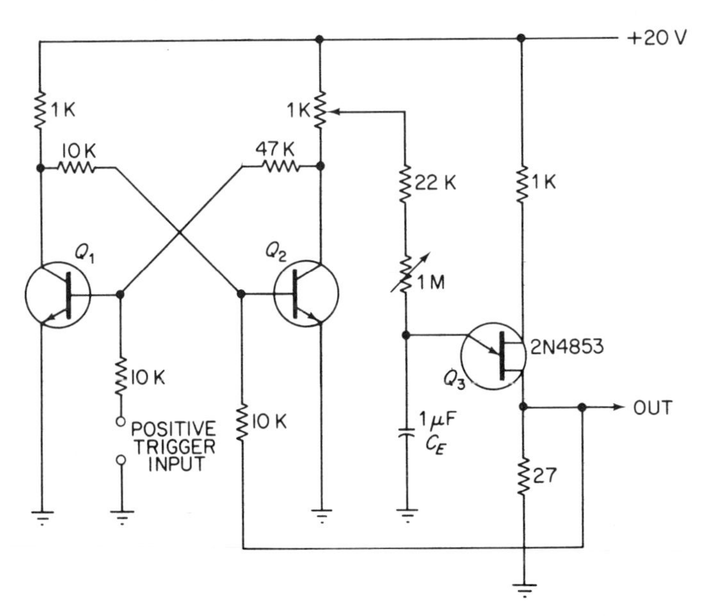

Fig. 1-31 UJT monostable circuit insensitive to change of bias voltage (Courtesy of Motorola Inc., Semiconductor Products Division)

The signal at the base of Q_2 triggers the flip-flop and Q_2 turns on. This leaves only a very small voltage applied to C_E, until the start of the next cycle. The voltage on C_E is the same at the start of the cycle, no matter how long a time has elapsed since the last cycle was completed. The circuit of Fig. 1-31 is nearly insensitive to changes in the supply voltage, and a change in V_1 from 10 V to 30 V results in a variation of approximately 2 per cent in timing. The output signal can also be taken from the flip-flop, in which case the output should be a clean square wave.

The hybrid one-shots offer the following advantages over conventional transistor one-shots: (1) The time delay available can generally be made longer; (2) the circuit can be triggered again immediately after the completion of one cycle; and (3) the circuit of Fig. 1-31 is relatively independent of changes in biased voltage, contrary to the behavior of conventional one-shots.

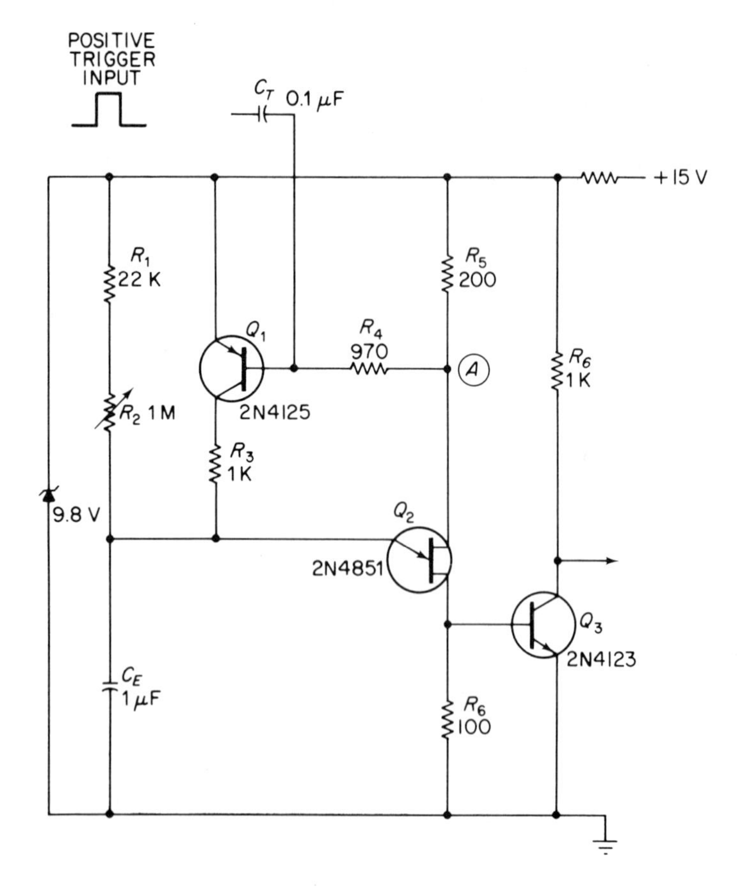

Fig. 1-32 UJT one-shot circuit (Courtesy of Motorola Inc., Semiconductor Products Division)

Hybrid One-Shot Circuits

By using a UJT and conventional transistors, it is possible to design a hybrid one-shot circuit that is not a true multivibrator. Figures 1-32 and 1-33 show such circuits.

In the circuit of Fig. 1-32, the UJT is normally ON and the emitter saturation current is supplied by transistor Q_1 which is also ON. When a positive trigger pulse is applied to the base of Q_1, Q_1 is turned OFF and, since the emitter current of Q_3 becomes less than I_v, the UJT turns OFF also. This starts the timing cycle. Capacitor C_E starts charging from the same voltage every time (namely, the saturation voltage of the UJT). Capacitor C_E charges through R_1 and R_2 and, when the voltage on C_E equals V_p, the UJT fires. This causes a drop in the voltage at point A, and transistor Q_1 turns ON, supplying the emitter current required to keep the UJT ON. Transistor Q_3 serves as an output device.

The output pulse period is determined by the value of C_E, R_1, and R_2. With R_1 fixed at 22 kΩ, and R_2 variable up to 1 megohm, the output pulse period is adjustable from about 20 milliseconds to 1 second. Different pulse periods can be obtained by changing the values of C_E (or R_1 and R_2, if necessary). The trigger should be 1 μS or longer for satisfactory operation.

Note that the circuit of Fig. 1-32 is voltage sensitive. Therefore, a Zener is used for voltage regulation.

The circuit of Fig. 1-33 is far simpler, but less stable and accurate, than the circuit of Fig. 1-32.

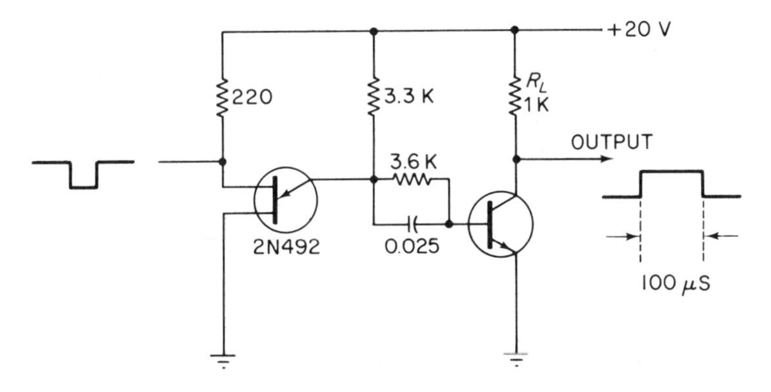

Fig. 1-33 Alternate simplified UJT one-shot circuit (Courtesy of General Electric Semiconductor Products Department, Syracuse, N.Y.)

1-6 Unijunction Time Delay Circuits

Because the ON and OFF times of a UJT are set by RC time constants, the UJT makes an excellent basic element for time delay circuits. In such timing circuits, the UJT is often used with a thyristor (typically an SCR or semiconductor controlled rectifier). However, the UJT need not be used with a thyristor, as is shown later in this section.

1-6-1 Basic UJT *Time Delay Circuit*

Figure 1-34 shows a typical time delay circuit that uses a UJT together with a low current SCR. The timing interval is started when power is applied to the circuit. At the end of the timing interval, which is determined by the values of R_T and C_T, the UJT fires the SCR, and the full supply voltage (minus about 1 V) is applied to the load.

The values shown in Fig. 1-34 are chosen on the basis of considerations described in Secs. 1-1 and 1-2. By suitable choice of R_T and C_T, the circuit can provide time delays from 0.4 millisecond to 1 minute. Load currents are limited by the SCR ratings. By using a precision calibrated resistor, such as a helipot, in place of R_T, the time delay can be set accurately over a wide range, after one initial calibration.

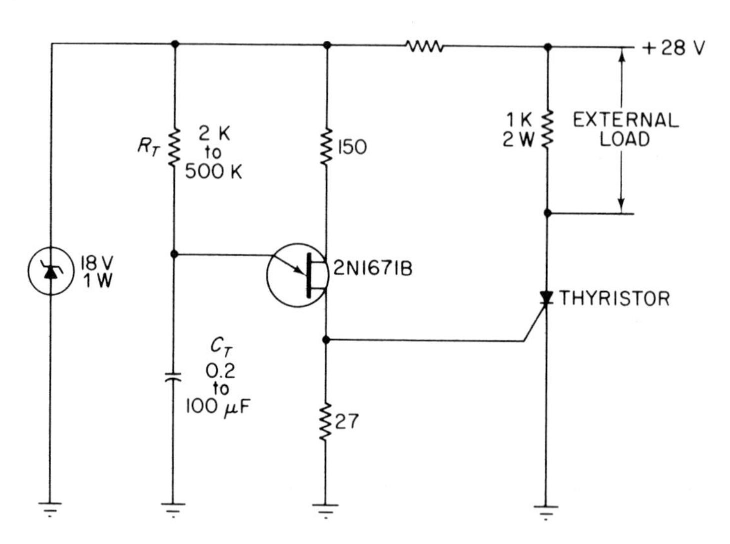

Fig. 1-34 Basic UJT time delay circuit (Courtesy of General Electric Semiconductor Products Department, Syracuse, N.Y.)

1-6-2 Practical UJT *Time Delay Circuit (with Relay)*

Figure 1-35 shows the basic UJT time delay circuit used in conjunction with a relay. The circuit values are determined on the basis of discussions in Secs. 1-1 and 1-2. With the variable resistance set at the maximum value, the

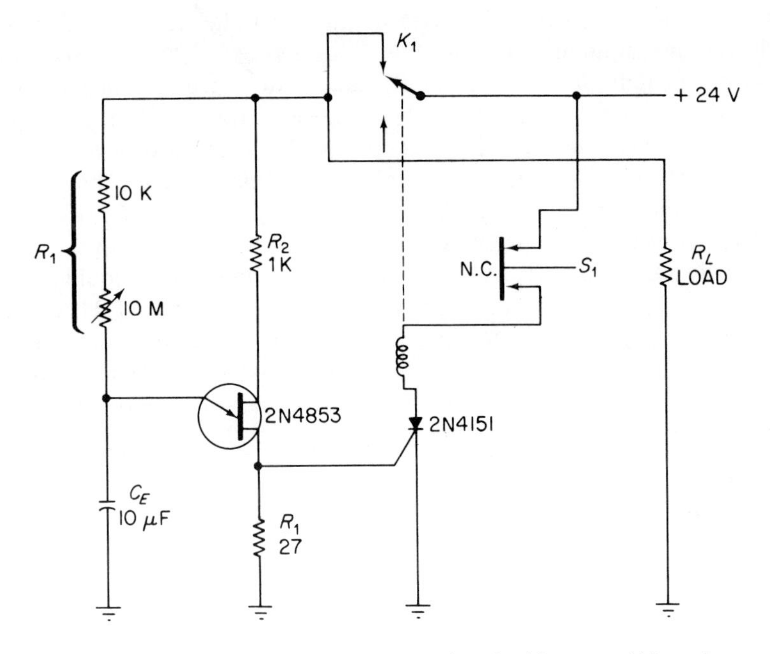

Fig. 1-35 Practical UJT time delay circuit with relay (Courtesy of Motorola Inc., Semiconductor Products Division)

time delay is approximately 160 seconds (assuming a standoff ratio of 0.8 for the UJT).

After the first cycle, the relay will normally be energized. When push button S_1 (normally closed) is pressed, the SCR turns OFF, the relay is de-energized, and power is applied to the relaxation oscillator as well as to the load. After a time delay varying from less than a second to approximately 160 seconds (as determined by the setting of the 10-megohm potentiometer), the UJT fires and turns the SCR ON. The relay is energized and power is removed from the UJT and load. Relay K_1 stays energized until S_1 is pushed again.

1-6-3 UJT *Circuits with Long Time Delays*

The time delay of a UJT oscillator is determined by the charge time of the capacitor. In order to have long time delays, R_E or C_E, or both, must be large. This problem brings up two tradeoffs.

If accuracy and repeatability of the oscillator waveform are required, the capacitor must have a leakage current that is *much smaller* than the charge current. A Mylar-type capacitor is usually good for this purpose. However, since Mylar capacitors are fairly expensive for large values of capacitance, it is usually preferable to increase R_E (rather than C_E) when long time constants (or delays) are required.

Unfortunately, large values of R_E can create a problem in relation to UJT peak-point emitter current I_p. When C_E is charged almost to the peak point, only a *small voltage* will appear across R_E. If R_E is very large, only a small current will flow. If the peak current of the UJT is large, the UJT will never fire if the current through R_E is not sufficient to supply I_p.

This problem can be partially overcome by using a UJT with a very low I_p (typically a fraction of 1 μA). For example, an annular UJT could have an I_p of about 0.2 μA, with an interbase voltage of 25 V. A typical UJT will have an I_p of 3 to 5 μA under maximum conditions. Also, when charging C_E through a large value of R_E, the charge current is initially large enough to accommodate I_p, but drops to a low value when the voltage across C_E is near the UJT peak voltage V_p.

For this reason, when it is not practical to use a UJT with a very low I_p and long time delays are required, it is preferable to charge C_E through a JFET (junction field effect transistor; refer to Chapter 2). Such a circuit is shown in Fig. 1-36. All of the values for this circuit are found by using the basic equations as discussed in Secs. 1-1 and 1-2. However, the charging current is controlled by the JFET.

Since the JFET is fully ON when there is no voltage from gate to source, the 10-megohm resistor R_3 determines the amount of OFF bias applied to the JFET. A constant current of less than 1 μA can easily be obtained with al-

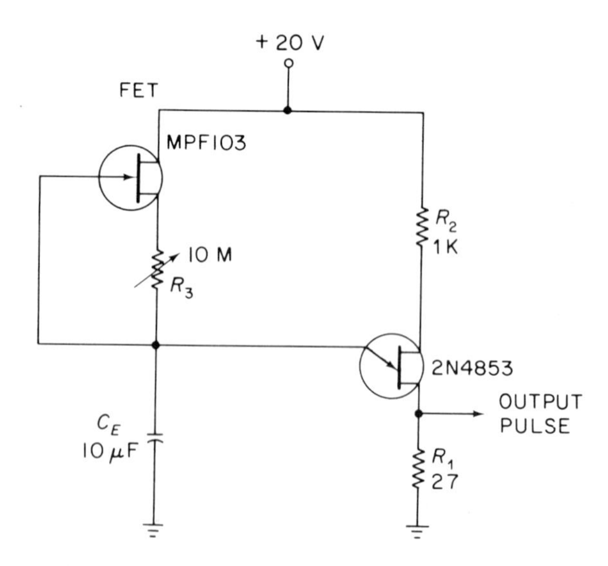

Fig. 1-36 UJT time delay circuit featuring constant current charging (Courtesy of Motorola Inc., Semiconductor Products Division)

most any type of UJT. This results in UJT oscillator circuits with time delays up to about 10 minutes.

When the capacitor is charged with a linear current, the charge time can be estimated from this equation:

$$t_{\text{charge}} \approx \frac{(V_p - V_v) \cdot C_E}{I_{\text{charge}}}$$

When C_E is in microfarads and I_{charge} (set by adjustment of R_3) is in microamperes, t will be in seconds. For example, if V_p is 14 V, V_v is 2 V, C_E is 10 μF, and I_{charge} is adjusted for 0.2 μA, the time is 10 minutes:

$$\frac{(14 - 2) \times 10}{0.2} = \frac{120}{0.2} = 600 \text{ seconds, or 10 minutes}$$

If time delays longer than 10 minutes are required, an emitter peak current as low as 0.2 μA can be objectionable. This can be overcome by supplying the charging current and the peak current from separate sources. Such a circuit is shown in Fig. 1-37.

In this circuit, transistor Q_1 and resistors R_1, R_2, and R_3 form a constant current source which can be adjusted to a few nanoamperes. Of course, the charge current is not sufficient to fire the UJT, even with an I_p of 0.2 μA. Thus, the peak current is supplied from another source. Transistor Q_2, acting as a source follower, supplies the current flowing into the emitter prior to firing. The transistor connected as diode D_1 provides a low impedance discharge path for C_E. D_1 must be selected to have a leakage *much lower* than the charge current.

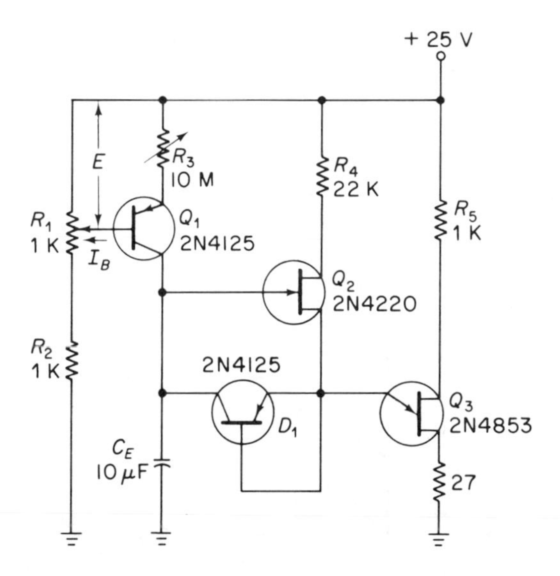

Fig. 1-37 UJT circuit for long time delays (Courtesy of Motorola Inc., Semiconductor Products Division)

The charge current to C_E is given as:

$$\text{charge} = \frac{E - V_{BE}}{R_3} - I_B$$

Since the I_B of Q_1 is typically small in relation to the I_{charge}, the charge time varies almost linearly with R_3. The voltage E, applied across R_3 and the base emitter junction of Q_1, is set by the variable resistor R_1. Time delays up to 10 hours are possible with this circuit. Resistor R_4 in series with the FET drain terminal must be large enough to prevent currents in excess of I_v from flowing when the UJT is ON. Should such currents flow, the UJT will not turn OFF, and the circuit will latch up.

1-7 Unijunction Multivibrators

In Sec. 1-5 we discussed circuits that use a UJT as the trigger source for hybrid multivibrators. It is possible to use a single UJT as a multivibrator. The basic UJT multivibrator circuit is shown in Fig. 1-38. Note that the circuit is essentially the same as for the basic relaxation oscillator, except for the addition of R_2 and CR_1. Diode CR_1 is forward-biased when C is being charged. Charging time is set by R_1 and C in the normal manner. However,

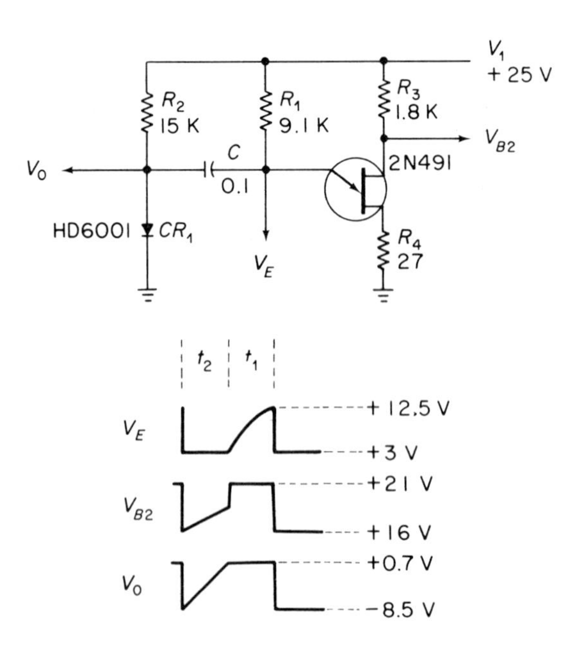

Fig. 1-38 UJT multivibrator (Courtesy of General Electric Semiconductor Products Department, Syracuse, N.Y.)

when C is discharging, CR_1 is reverse-biased (the anode of the diode is negative). As long as the capacitor is charged, the discharge current must flow through R_2, with the R_2C constant determining the discharge time.

As shown by the waveforms, the output at Base-2 is an approximate square wave, with the ON and OFF intervals separately controlled by R_1 and R_2. The time t_1 is the period for which the UJT is OFF, and t_2 is the period for which the UJT is ON and CR_1 is reverse-biased.

A fairly accurate approximation of these times can be made by using these equations:

$$t_1 \approx R_1 C \ln \left[\frac{V_1 - V_E}{V_1 - V_p}\right]$$

$$t_2 \approx R_2 C \ln \left[\frac{V_1 + V_p - V_E}{V_1}\right]$$

where V_E is the emitter voltage measured at an emitter current $I_E = V_1(R_1 + R_2)/R_1R_2$ and may be taken from the emitter characteristic curves.

However, a simplified rule is to *make R_2 approximately twice the value of R_1* when t_1 is to equal t_2.

In a practical experimental circuit, calculate trial values for a given t_1 on the basis of a simple R_1C time constant. Then use a variable resistance for R_2, starting with a value twice that of R_1.

The basic UJT multivibrator may be coupled to a conventional transistor by means of a circuit such as shown in Fig. 1-39. In this circuit, the emitter-to-base diode of the transistor takes the place of the diode in the circuit of Fig. 1-38. The Fig. 1-39 circuit has the advantage that the load is completely isolated from the timing portion of the circuit. However, all of the timing values remain the same. Note that in the circuit of Fig. 1-39, both Base 1 and Base 2 resistors are omitted. The Base 1 resistor is actually not required in either circuit, although the addition of a Base 1 resistor to either circuit will minimize excessive emitter current when the UJT fires (if that is a prob-

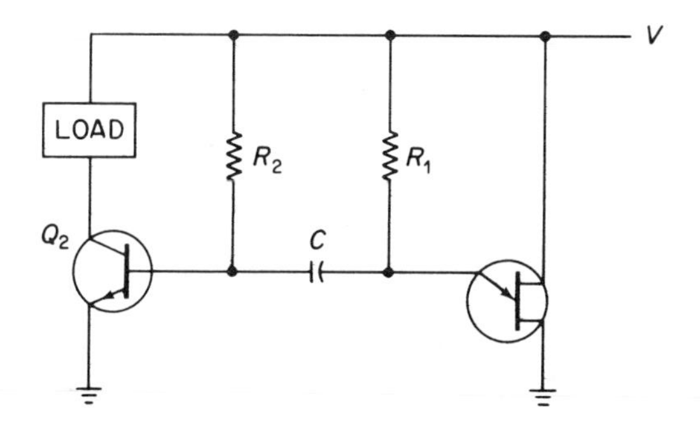

Fig. 1-39 UJT multivibrator coupled to load through an *NPN* transistor (Courtesy of General Electric Semiconductor Products Department, Syracuse, N.Y.)

lem with any particular application). The Base 2 resistor provides temperature compensation, as described in Sec. 1-1, and can be used or omitted to fit a particular application. In the case of the Fig. 1-38 circuit, the Base 2 resistor R_3 is required since the output is developed across this resistor. In the Fig. 1-39 circuit, output is taken from transistor Q_2, and no temperature compensation is provided.

1-8 Unijunction Frequency Dividers

The circuit of Fig. 1-40 illustrates how the simple UJT relaxation oscillator can be adapted to make a simple frequency divider. The Base 2 current pulses (generated when one UJT fires) generate a negative pulse across the 120-ohm resistors. In turn, these pulses synchronize the next UJT oscillator. This circuit illustrates the use of both emitter synchronization for the first UJT and Base 2 synchronization for the second and third UJTs. In a practical application, it is necessary to use potentiometer adjustments for the capacitor charging resistors if division ratios higher than two are required.

The frequency divider of Fig. 1-40 has the disadvantage of working over a very narrow range of input frequency. A circuit that gives a constant division ratio over a much wider range of input frequencies is shown in Fig. 1-41. The circuit shown is a single stage of a divider that may have an unlimited number of stages.

In the circuit of Fig. 1-41, input consists of positive pulses of arbitrary separation at the base of the NPN transistor. During each pulse, capacitor C_1 is discharged at the transistor and the diode D_1. At the end of the pulse, C_1 is charged through R_1, and an equivalent charge flows into C_2. After a certain number of pulses, C_2 is charged to the peak-point voltage of the UJT

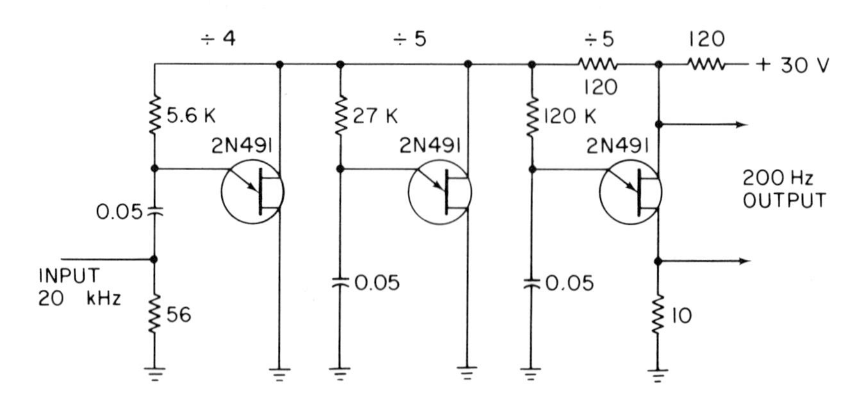

Fig. 1-40 UJT frequency divider, providing 100-to-1 frequency division (Courtesy of General Electric Semiconductor Products Department, Syracuse, N.Y.)

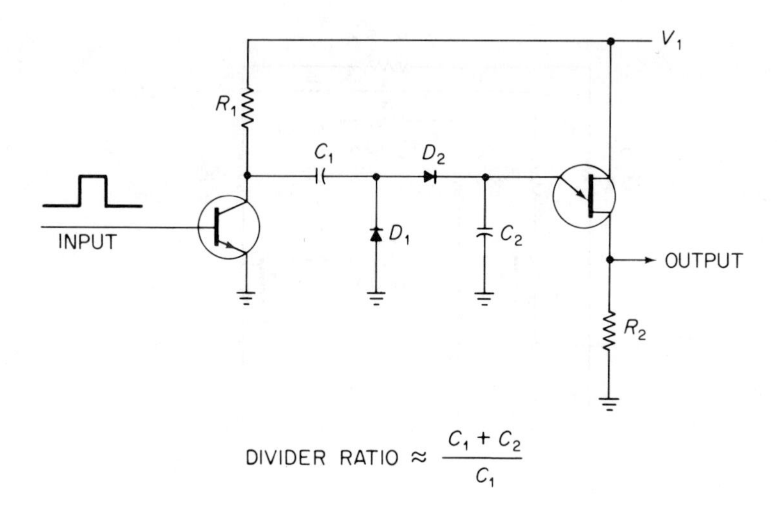

Fig. 1-41 Single stage UJT frequency divider (Courtesy of General Electric Semiconductor Products Department, Syracuse, N.Y.)

so that the UJT fires and discharges C_2, thus ending the cycle. The positive pulse across R_2 can serve as an input for the next divider stage. The approximate divider ratio is $(C_1 + C_2)/C_1$. This circuit has the disadvantage of requiring more components than the simple circuit of Fig. 1-40. However, the circuit of Fig. 1-41 has the advantage of working with a much wider range of frequencies.

1-9 Unijunction Variable Phase Generators

The basic UJT relaxation oscillator can readily be adapted to provide two pulses alternating from two separate outputs. Such a circuit is shown in Fig. 1-42.

The circuit consists of two basic relaxation oscillators, coupled by capacitor C_3 in such a way as to achieve sychronization, as well as with a phase shift between the two outputs. The frequency and phase shift are varied independently. Potentiometer R_1 controls frequency, whereas R_2 sets the phase shift.

The values shown in Fig. 1-42 provide for a frequency range (R_1) of about 200 to 800 Hz, with a phase shift (R_2) of about 70° to 290°. As usual, the frequency of each oscillator is set by the RC time constant, as described in Sec. 1-1. Likewise, the Base 1 and Base 2 resistor values are set by output and temperature stabilization factors discussed in Sec. 1-1. As a first trial, the values of R_2 and R_3 should be equal to the emitter charging resistors, while R_1 should be one-half the same value.

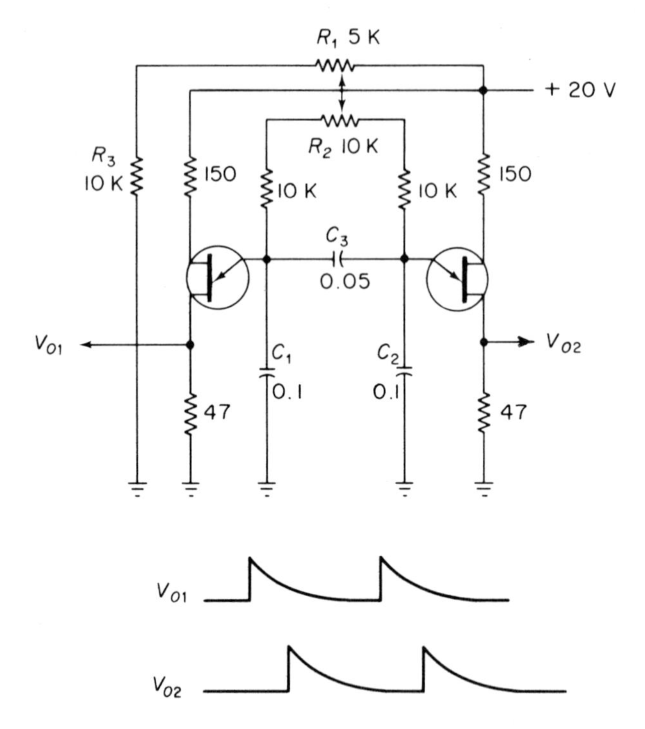

Fig. 1-42 UJT variable phase generator (Courtesy of General Electric Semiconductor Products Department, Syracuse, N.Y.)

The range of phase angle control is set by the ratio of C_3 to C_1 and C_2. The phase angle range is increased when C_3 is decreased relative to C_1 and C_2. Thus, in practical design, the values of C_1 and C_2 are selected first. Then a proportional value of C_3 is selected to give the desired phase angle control range.

The output pulse amplitude for the circuit shown is about 4 V peak. The approximate pulse width is 10 μS.

1-10 Unijunction Voltage Sensing Circuit

The high sensitivity of the UJT and extreme stability of its firing characteristic make the UJT ideally suited for use in voltage-sensing circuits. Such a circuit is shown in Fig. 1-43. Note that this circuit includes a simple *floating* power supply with Zener diode regulation which operates from a 115-V, 60-Hz, a-c line. Another power supply in the 20- to 24-V range can be substituted, provided that the supply is floating (with neither positive nor negative at ground).

A reference voltage is applied to the terminal shown, and input voltage

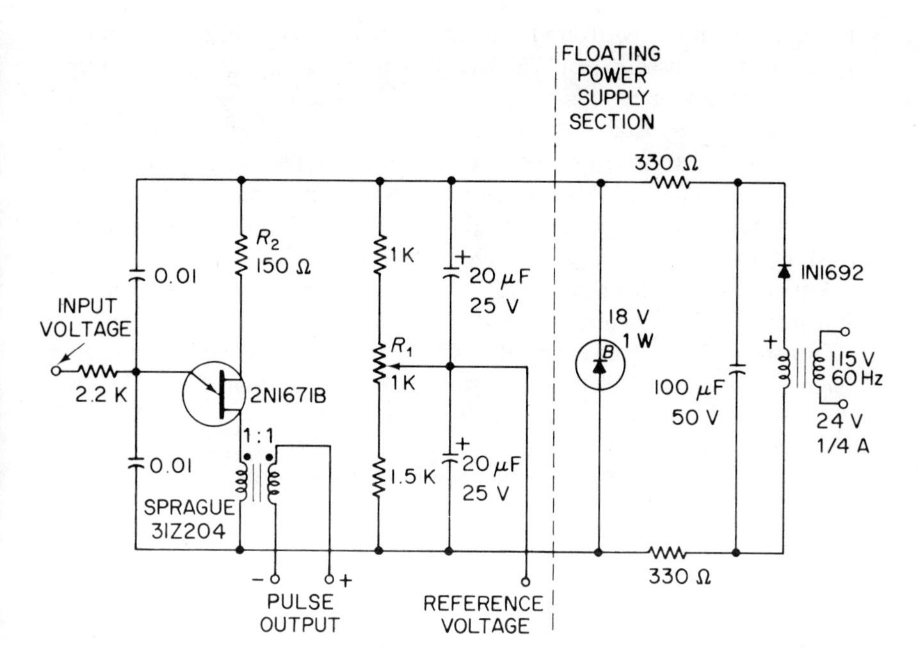

Fig. 1-43 UJT voltage-sensing circuit (Courtesy of General Electric Semiconductor Products Department, Syracuse, N.Y.)

(to be compared against the reference) is applied to the other terminal (UJT emitter). Whenever input voltage exceeds reference voltage, the UJT fires and produces an output pulse. This pulse can be used to fire an SCR or other pulse-sensitive device.

The values of R_1 and its two series resistors are chosen to provide an approximate 7-V spread for the trigger level. The value of R_2 is chosen to provide for temperature compensation.

Using the values shown, a typical trigger current of less than 5 μA is required. The short-term stability is about 1 mV (or better), while long-term voltage stability is 10 mV or better.

1-11 Unijunction Ring Counters

A ring counter consists of a number of *bistable circuits* that are connected to form a closed ring. In the normal operating condition, all the bistable circuits (stages) of the ring counter are biased in the same state, except for one stage which is biased in the opposite state or ON state. Connections between the stages are arranged so that when a trigger pulse from an external source is applied to all of the stages in parallel, the ON state is advanced one stage (either to the right or left). If the ring counter has n stages, the ON state will advance completely around the ring after n trigger pulses are applied. Ring counters are used in applications such as frequency dividers, counters, and sequential gating circuits.

The basic UJT bistable circuit (described in Sec. 1-4 and shown in Fig. 1-21) can be used for each stage of a ring counter. The basic UJT bistable circuit requires only one UJT, one diode, two resistors, and one capacitor. This compares favorably with a typical transistor flip-flop which requires two transistors, two diodes, six resistors, and three capacitors.

A typical four-stage ring counter that uses basic UJT bistable elements is shown in Fig. 1-44. Operation of the circuit is as follows.

Transistor Q_1 (a conventional junction transistor) is used as a trigger amplifier. Assume that UJT Q_2 is ON. The collector current from Q_1 flows through R_3, D_2, and the emitter of Q_2. Voltage at the emitter of Q_2 is V_B (approximately 3 V), while voltage at the emitters of Q_3, Q_4, and Q_5 is equal to V_1 (10 V) since each of these transistors is in the OFF state.

When a trigger pulse is applied, Q_1 is cut off and voltage at collector Q_1 falls to zero. The emitter current of Q_2 then drops to zero, and Q_2 turns OFF. During the trigger pulse, capacitor C_5 discharges through R_9, R_3, and D_2; however, capacitor C_2 will retain its voltage ($V_1 - V_B = 7$ V) since C_2 is prevented from discharging by the reverse-biased diode D_3 and the cut-off emitter of Q_3.

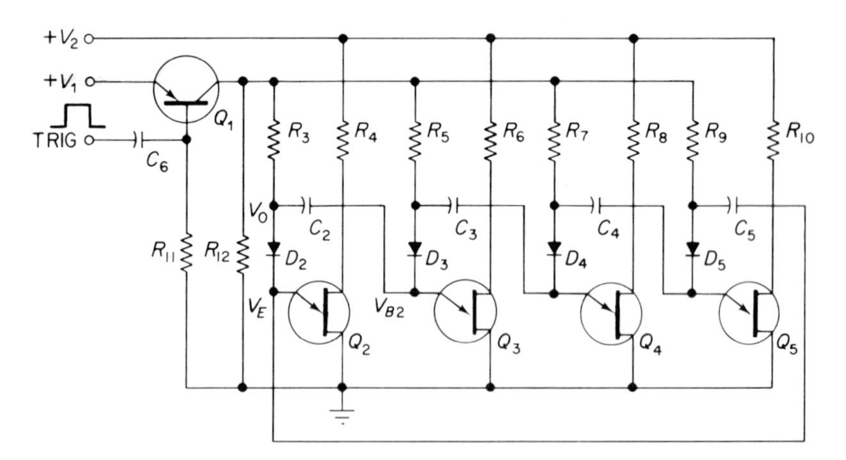

CIRCUIT VALUES:

V_1 = 10.0 VOLTS ± 1%
V_2 = 20.5 VOLTS ± 1%
TRIG > 5 VOLTS, 10 − 30 μS
$D_2 - D_5$ = HD6001 (SILICON)
Q_1 − 2N395
$Q_2 - Q_5$ = 2N492

$R_1 = R_2 = R_3 = R_{10}$ = 470 Ω ± 5%
R_{11} = 5.6 K ± 10%
R_{12} = 2.2 K ± 10%
$C_1 = C_2 - C_5$ = 1000 pF ± 20%
C_6 = 0.1 μF ± 20%

Fig. 1-44 Basic UJT ring counter (Courtesy of General Electric Semiconductor Products Department, Syracuse, N.Y.)

At the end of the trigger pulse, voltage at the collector of Q_1 rises to V_1. Since there is no voltage on C_3, C_4, and C_5, voltages at the emitters of Q_2, Q_4, and Q_5 also rise to V_1. As peak-point voltage at each of these transistors is greater than V_1, none of these transistors turn ON; however, the voltage on C_2 is equal to $V_1 - V_B$ so that voltage at the emitter of Q_3 rises to $2V_1 - V_B$ (about 17 V), which is greater than V_p. This causes Q_3 to turn ON. Note that diode D_3 blocks that current from C_2 and causes all of it to flow into the emitter of Q_3. This prevents C_2 from triggering Q_4 as well as Q_3.

In similar manner, each successive trigger pulse advances the ON state one stage to the right from Q_3 to Q_4 to Q_5, and then back to Q_2.

1-11-1 Circuit Features

The ring counter circuit shown in Fig. 1-44 has a number of advantageous features:

(1) the circuit can be designed to work with the normal range of parameters for a standard type of UJT;

(2) the coupling capacitor cannot discharge during the trigger pulse so that the trigger pulse width can vary over a wide range;

(3) triggering takes place at the emitter where the required trigger amplitude is lowest (compared to Base 1 or Base 2 triggering) so that the operating margins are maximum;

(4) the diodes prevent the turn-on trigger pulse from being applied to more than one stage in the forward direction;

(5) separate loads are used in each stage (R_3, R_5, R_7, R_9);

(6) small resistors (R_4, R_6, R_8, R_{10}) may be used in series with Base 2 for stabilization of the peak-point voltage (outputs may be taken from these resistors, if desired);

(7) negative pulses at the Base 2 terminals may be used for initially turning ON one stage.

1-11-2 Design Considerations

Several design considerations must be made for the circuit of Fig. 1-44:

1. The principal design conditions are concerned with the allowable range of peak-point voltage, and V_B. If $V_{p(\min)}$, $V_{p(\max)}$, and $V_{B(\max)}$ are the exterme values found for the worst conditions of η, R_{BB}, and temperature, respectively, these conditions may be written:

$$V_{1(\min)} < V_{p(\min)} \approx \eta_{\min} V_{2(\min)}$$

$$2V_{1(\min)} - V_{B(\max)} > V_{p(\max)} \approx \eta_{\max} V_{2(\max)}$$

Definitions for these terms are shown in Fig. 1-21 and discussed in Sec. 1-4.

2. *The minimum pulse width* is determined by the time constant $2R_1C_1$. The pulse width should be *greater than three times the* $2R_1C_1$ constant to permit the capacitor coupling to the preceding stage to discharge during the pulse interval. *The maximum allowable pulse width* is determined by the value of C_1 and the leakage currents of the diodes and emitter of the UJT, at the maximum operating temperature. Assuming typical diode and UJT emitter leakage, the maximum pulse width should be about 100 times the $2R_1C_1$ constant, although the circuit could operate with trigger widths up to 1500 times the $2R_1C_1$ constant.

3. The resistor R_2 used in series with Base-2 provides temperature compensation for the peak-point voltage. The optimum value of R_2 is based on the conditions described in Sec. 1-1.

4. The value of resistor R_1 may be chosen over a wide range. The upper limit is set by the emitter characteristics, since R_1 must be small enough to produce good voltage swing at the emitter ($V_1 - V_B$). If capacitive loads are connected at the emitters of the UJT, the ON state (point B, Fig. 1-21) must fall to the right of the valley point; otherwise, the circuit will be unstable and possibly operate as a relaxation oscillator rather than as a bistable element. The lower limit on R_1 is set by the allowable power dissipation of the UJT. The use of resistors between the emitter and ground is not advisable, since they will reduce triggering margins.

5. The voltages V_1 and V_2 should be stable within ± 1 per cent. However, if V_1 and V_2 are derived from a common reference, they can vary over a much larger range as long as their ratio is constant within ± 1 per cent.

6. Design margins can generally be improved by increasing V_1 and V_2, since V_P will increase by a greater amount than V_B.

7. A germanium transistor can be used for Q_1 if the maximum ambient temperature is below 25°C. At higher temperatures a silicon transistor must be used.

8. Typical waveforms for the circuit of Fig. 1-44 are shown in Fig. 1-45.

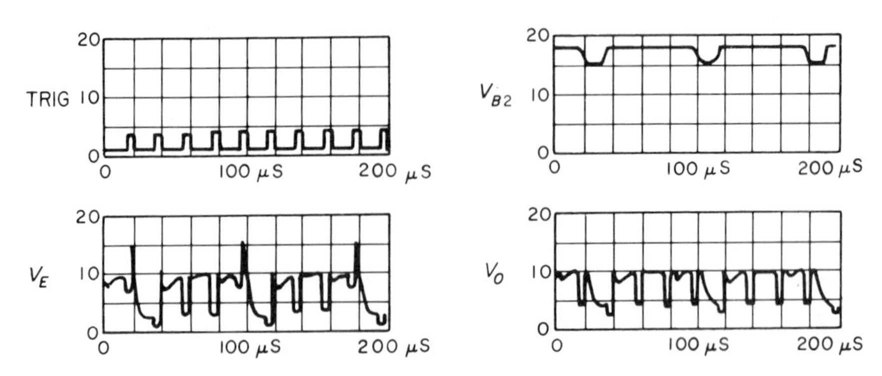

Fig. 1-45 Typical waveforms for basic UJT ring counter (Courtesy of General Electric Semiconductor Products Department, Syracuse, N.Y.)

1-11-3 Circuit Variations

Alternative methods of triggering are shown in Fig. 1-46. These methods have the advantage of producing less noise on the resistors R_3, R_5, R_7, R_9, as compared to the circuit of Fig. 1-44.

Negative trigger pulses applied at the Base 2 terminals can be used to turn on any of the UJTs. Negative triggers applied at Base 2 during the main trigger pulse can also be used to eliminate the ON state at a particular stage in the ring counter. It is also possible to have two or more UJTs in the ON state at the same time. The ring counter circuit shown in Fig. 1-44 thus has the essential features of a shift register. (The design and application of shift registers is described in the author's *Handbook of Logic Circuits*, Reston Publishing Company, Inc.).

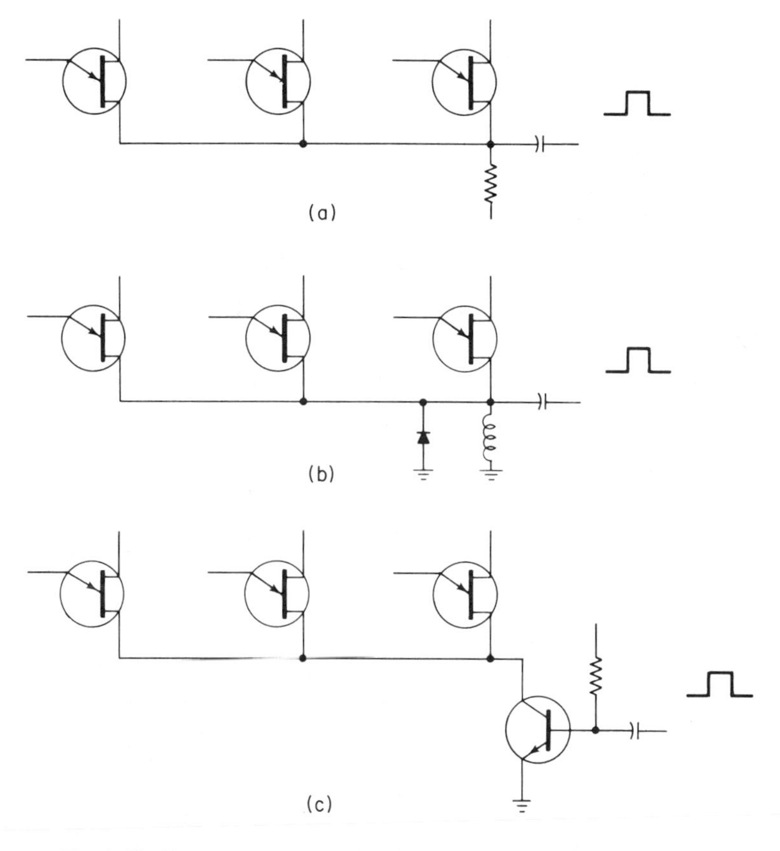

Fig. 1-46 Alternate base-one triggering method for UJT ring counter (Courtesy of General Electric Semiconductor Products Department, Syracuse, N.Y.)

The methods shown in Fig. 1-46 also permit the resistors to be connected to a common ground point. In a circuit with a large number of stages, however, the sum of the interbase currents can cause the current flowing through the trigger circuit to be quite large. Note that the basic UJT relaxation oscillator of Sec. 1-1, when used as a pulse generator, makes an excellent trigger source for the ring counter circuit.

1-12 Unijunction Regenerative Amplifier

The basic UJT relaxation oscillator may be adapted to form a regenerative pulse amplifier. Such an amplifier is shown in Fig. 1-47. Note that the circuit is essentially the basic relaxation oscillator with two resistors added.

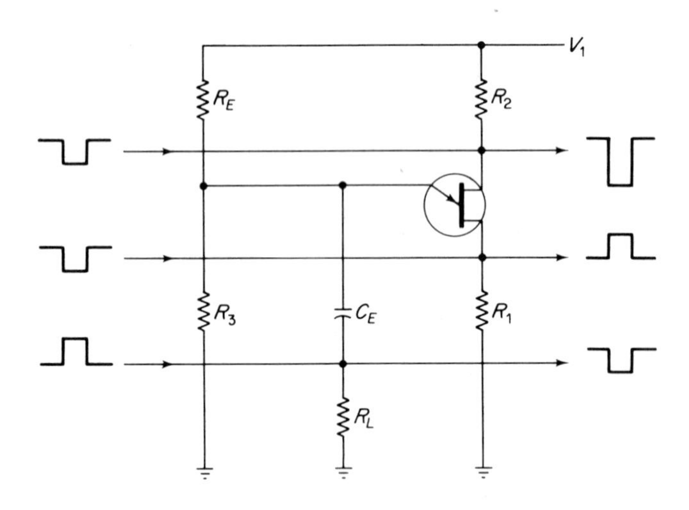

$R_L \approx Z$ OF TRIGGER, OR Z OF OUTPUT LOAD

$$R_3 \approx \frac{R_E}{V_1 - V_{BIAS}} \times V_{BIAS}$$

$$R_E \gg \frac{V_1 - V_V}{I_V} \qquad V_{BIAS} = V_P - 0.5 \text{ TRIGGER}$$

$$C_E \approx \frac{\text{PERIOD}}{RE \times \ln \frac{1}{1-\text{STANDOFF RATIO}}}$$

Fig. 1-47 Basic UJT regenerative amplifier (Courtesy of General Electric Semiconductor Products Department, Syracuse, N.Y.)

Resistor R_L provides a load for a trigger input or a pulse output. Resistor R_3 forms a voltage divider with resistor R_E. The voltage at the junction of R_3 and R_E is set so that the UJT will not fire except in the presence of a trigger. That is, R_3 and R_E have a ratio such that the emitter voltage does not exceed the peak-point (firing) voltage for the quiescent state.

As shown in Fig. 1-47, inputs can be applied to any of three points (Base 1, Base 2, or emitter). Likewise, outputs can be taken from any of these points. In operation, a trigger pulse is applied to one of the inputs, and a corresponding trigger output (in greatly amplified form) is taken from another of the points. For example, a trigger of 0.1 V can be applied to the emitter, with a 7-V output taken at Base 1.

1-12-1 Design Considerations

The values of R_1, R_2, and C_E are selected on the same basis as discussed for the UJT relaxation oscillator in Sec. 1-1. The time constant for R_E and C_E should be approximately the same as the period of the trigger.

The value of R_E is found in the same way essentially as for the basic oscillator, but with a minor difference. The maximum value of R_E need not be of great concern in the regenerative amplifier. The UJT is held in the OFF condition until a trigger is applied by the voltage drop across R_3 (or by the voltage division at the jumction of R_E and R_3). Thus, no calculation need be made for $R_{E(\max)}$. The minimum value for R_E is found in the same way as for the basic oscillator; that is, the minimum R_E that can be used with the regenerative amplifier is set by

$$R_E > \frac{V_1 - V_v}{I_v} = R_{E(\min)}$$

In order to assure turn-off after the trigger pulse is removed, however, the value of R_E should be two to three times larger than $R_{E(\min)}$.

The value of R_L is selected to match the load impedance of the input trigger source, or the output, depending upon the application.

The value of R_3 is set by the desired bias voltage. The ratio of R_E and R_3 sets the fixed bias that must be overcome by the input trigger. V_{bias} is set at V_p (peak voltage of the UJT), less one-half the input trigger.

1-12-2 Design Example

Assume that the circuit of Fig. 1-47 is to provide an output of 3 V (minimum) across R_1, with a trigger input of 0.4 V from a source of 50 ohms. The trigger frequency is in the 200- to 300-Hz range. The available source voltage V_1 is 20 V. A test of the UJT (or data-sheet information) shows a typical peak voltage V_p of 14 V, valley voltage V_v of 2 V, and valley current I_v of

6 mA. The standoff ratio is 0.8 and the interbase resistance R_{BB} is approximately 7 kΩ.

Design starts with R_E. Using the $R_{E(\min)}$ equation, the absolute minimum value is 3 kΩ (20 − 2 = 18; 18/0.006 = 3 kΩ). First trial value of R_E should be three times $R_{E(\min)}$ or 9 kΩ. Use 10 kΩ to simplify calculations.

With the value of R_E established at 10 kΩ, the value of C_E is selected to make the circuit frequency and trigger frequency about the same. Using the 300-Hz frequency for the first trial value, a period of about 3000 μS is required. With a period of 3000 μS, an R_E of 10 kΩ, and a standoff ratio of 0.8, the equation of Fig. 1-1 shows that C_E should be approximately 0.2 μF:

$$\frac{3000}{10{,}000 \times \ln 1/(1 - 0.8)} = \frac{3000}{15{,}000} = 0.2\ \mu\text{F}$$

With a value of 0.2 μF for C_E and a required minimum of 3 V peak output across R_1, Fig. 1-9 shows a value of 20 ohms for R_1.

The value of R_2 is found by using the equation of Fig. 1-8. With a source voltage of 20 V, a standoff ratio of 0.8, and an interbase resistance of 7 kΩ, the value of R_2 is 0.015 × 20 × 0.8 × 7000 = 1680 ohms.

The value of R_L should be equal to the input trigger source impedance of 50 ohms.

The value of R_3 is set by the relationship between R_E, V_1, and V_{bias}. With an input trigger of 0.4 V, V_{bias} is 0.5 × 0.4 = 0.2 V; 14 − 0.2 = 13.8 V. With R_E at 10 kΩ, V_1 at 20 V, and V_{bias} at 13.8 V, the first trial value of R_3 is 20 − 13.8 = 6.2 V; 10,000/6.2 = 1611; 1611 × 13.3 = 22,231. In a practical experimental circuit, R_3 should be adjusted to approximately 22,231. Then the circuit should be tested to make sure that the UJT does not trigger (start to operate as a relaxation oscillator) when no trigger is applied, but that the UJT does fire when the 0.4-V trigger is applied.

CHAPTER 2

Field Effect Transistor Circuits

Before going into specific design considerations for field effect transistor (FET) circuits, we shall discuss basic FET operating characteristics. This is necessary since FET characteristics are unique when compared to any other type of transistor. For example, a FET is often biased at the zero-temperature-coefficient point. This is an operating point where the FET drain-source current will not vary with temperature. Likewise, the characteristics shown on FET data sheets do not correspond to those of conventional two-junction transistors. It is necessary to analyze these data-sheet characteristics, as they apply to practical design. FET types and operating modes are discussed fully in the author's *Practical Semiconductor Databook for Electronic Engineers and Technicians*. The following paragraphs summarize this information.

2-1 Advantages and Disadvantages of FETs

The FET has several advantages over a conventional transistor. The FET is relatively free of noise, and is more resistant to the degrading effects of nuclear radiation because carrier lifetime effects are comparatively unimportant to FET operation. The FET is inherently more resistant to burnout than is a conventional two-junction (or bipolar) transistor.

CHARACTERISTICS	VACUUM TUBE	JFET	MOSFET	BIPOLAR
Input Impedance	High	High	Very High	Low
Noise	Low	Low	Unpredictable	Low
Warm-Up Time	Long	Short	Short	Short
Size	Large	Small	Small	Small
Power Consumption	Large	Small	Small	Small
Ageing	Noticeable	Not Noticeable	Noticeable	Not Noticeable
Bias Voltage Temp Coefficient	Low, Not Predictable	Low Predictable	High Not Predictable	Low Predictable
Typical Gate/Grid Current	1 nA	0.1 nA	10 pA	—
Gate/Grid Current Change with Temp	High Unpredictable	Medium Predictable	Low Unpredictable	—
Reliability	Low	High	High	High
Sensitivity To Overload	Very Good	Good	Poor	Good

Fig. 2-1 Comparison of FET characteristics (Courtesy Motorola)

There are additional advantages for certain design considerations. For example, the high input impedance (typically several megohms) is very useful in impedance transformations, chopper, and switching applications. Since the FET is a voltage-controlled device, the FET can readily be "self-biased." This frequently permits a more simple circuit than is possible with a bipolar transistor. The FET also has a nonlinear region of operation (Sec. 2-3) that can be used for automatic gain control applications.

The junction field effect transistor (or JFET, Sec. 2-2) has a very high output resistance, making it useful as a constant current source. Figure 2-1 illustrates a comparison of some JFET key parameters and their relative magnitudes to those of vacuum tubes, bipolar transistors, and MOSFETs.

When compared with bipolar transistors, the chief shortcoming of the FET is its relatively small gain-bandwidth product. Although the JFET is free from carrier-transit-time limitations, parasitic capacitances limit the FET at higher frequencies.

2-2 Types of FETs and Modes of Operation

There are two types of field effect transistors: Junction (JFET) and metal-oxide-silicon (MOSFET). As the names imply, a JFET uses the characteristics of a reverse-biased junction to control the drain-source current;

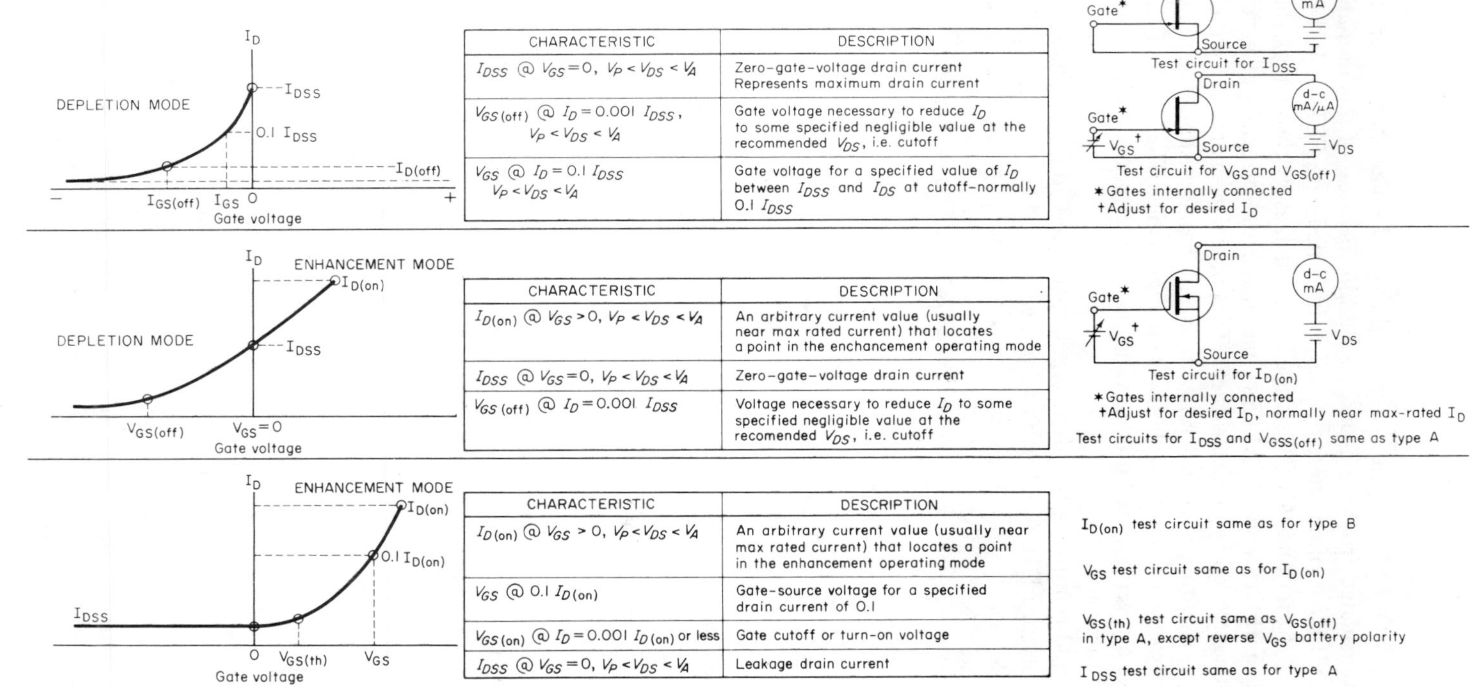

CHARACTERISTIC	DESCRIPTION
I_{DSS} @ $V_{GS} = 0$, $V_P < V_{DS} < V_A$	Zero-gate-voltage drain current Represents maximum drain current
$V_{GS(off)}$ @ $I_D = 0.001\ I_{DSS}$, $V_P < V_{DS} < V_A$	Gate voltage necessary to reduce I_D to some specified negligible value at the recommended V_{DS}, i.e. cutoff
V_{GS} @ $I_D = 0.1\ I_{DSS}$ $V_P < V_{DS} < V_A$	Gate voltage for a specified value of I_D between I_{DSS} and I_{DS} at cutoff-normally $0.1\ I_{DSS}$

CHARACTERISTIC	DESCRIPTION
$I_{D(on)}$ @ $V_{GS} > 0$, $V_P < V_{DS} < V_A$	An arbitrary current value (usually near max rated current) that locates a point in the enchancement operating mode
I_{DSS} @ $V_{GS} = 0$, $V_P < V_{DS} < V_A$	Zero-gate-voltage drain current
$V_{GS(off)}$ @ $I_D = 0.001\ I_{DSS}$	Voltage necessary to reduce I_D to some specified negligible value at the recomended V_{DS}, i.e. cutoff

CHARACTERISTIC	DESCRIPTION
$I_{D(on)}$ @ $V_{GS} > 0$, $V_P < V_{DS} < V_A$	An arbitrary current value (usually near max rated current) that locates a point in the enhancement operating mode
V_{GS} @ $0.1\ I_{D(on)}$	Gate-source voltage for a specified drain current of 0.1
$V_{GS(on)}$ @ $I_D = 0.001\ I_{D(on)}$ or less	Gate cutoff or turn-on voltage
I_{DSS} @ $V_{GS} = 0$, $V_P < V_{DS} < V_A$	Leakage drain current

Fig. 2-2 Static characteristics for three FET types (Courtesy Motorola)

with the MOSFET the gate is a metal deposited on an oxide layer. The gate is thus insulated from the source and drain. Because of this insulated gate, the MOSFET is often referred to as an insulated-gate FET, or IGFET. The terms MOSFET and IGFET are used interchangeably throughout this book.

Both JFETs and MOSFETs operate on the principle of a "channel" current controlled by an electric field. The control mechanisms for the two are different, resulting in considerably different characteristics. The main difference between the two is in the gate characteristics. The input of the JFET acts like a reverse-biased diode, while the input of a MOSFET is similar to a small capacitor.

In addition to the two basic types, there are two fundamental modes of operation for FETs—*depletion* and *enhancement*. These modes are illustrated in Fig. 2-2 which shows the transfer characteristics and basic test circuits for each mode.

In the depletion mode, maximum drain current (I_{DSS}) flows when the gate-source voltage (V_{GS}) is zero, and decreases for increasing V_{GS}. Enhancement mode is just the opposite, in that minimum drain current flows at $V_{GS} = 0$ but increases for increasing V_{GS}.

FETs designated as type A are defined in the depletion mode only. Type B FETs are defined in both the depletion and enhancement modes. Type C FETs operate in the enhancement mode only.

The test circuits in Fig. 2-2 show the biasing for *N*-channel FETs. Note that V_{DS} is always positive for the three *N*-channel types. In the useful range of operation, V_{GS} is negative for a type A FET, positive for type C, and either polarity for type B. For a *P*-channel FET, all polarities must be reversed.

2-3 Basic FET Operating Regions

The FET has three distinct characteristic regions, only two of which are operational. Figure 2-3a, the output transfer characteristic, illustrates the different regions. Below the pinch-off voltage V_P, the FET operates in the *ohmic* or *resistance region*. Above the pinch-off voltage, up to the drain-source breakdown voltage $V_{(BR)DSS}$, the FET operates in the *constant current region*. The third region, above the breakdown voltage, is the avalanche region where the FET is not operated.

The drain-source resistance r_{DS} at any point on these curves is given by the slope of the curve at that point. Above the pinch-off voltage, changes in the drain-source voltage V_{DS} result in small changes in drain current I_D. This produces a very high drain-source resistance, and is characteristic of a constant-current source. Also, the actual operating drain current is variable, and is dependent on the gate-source voltage. This results in a voltage-controlled current source.

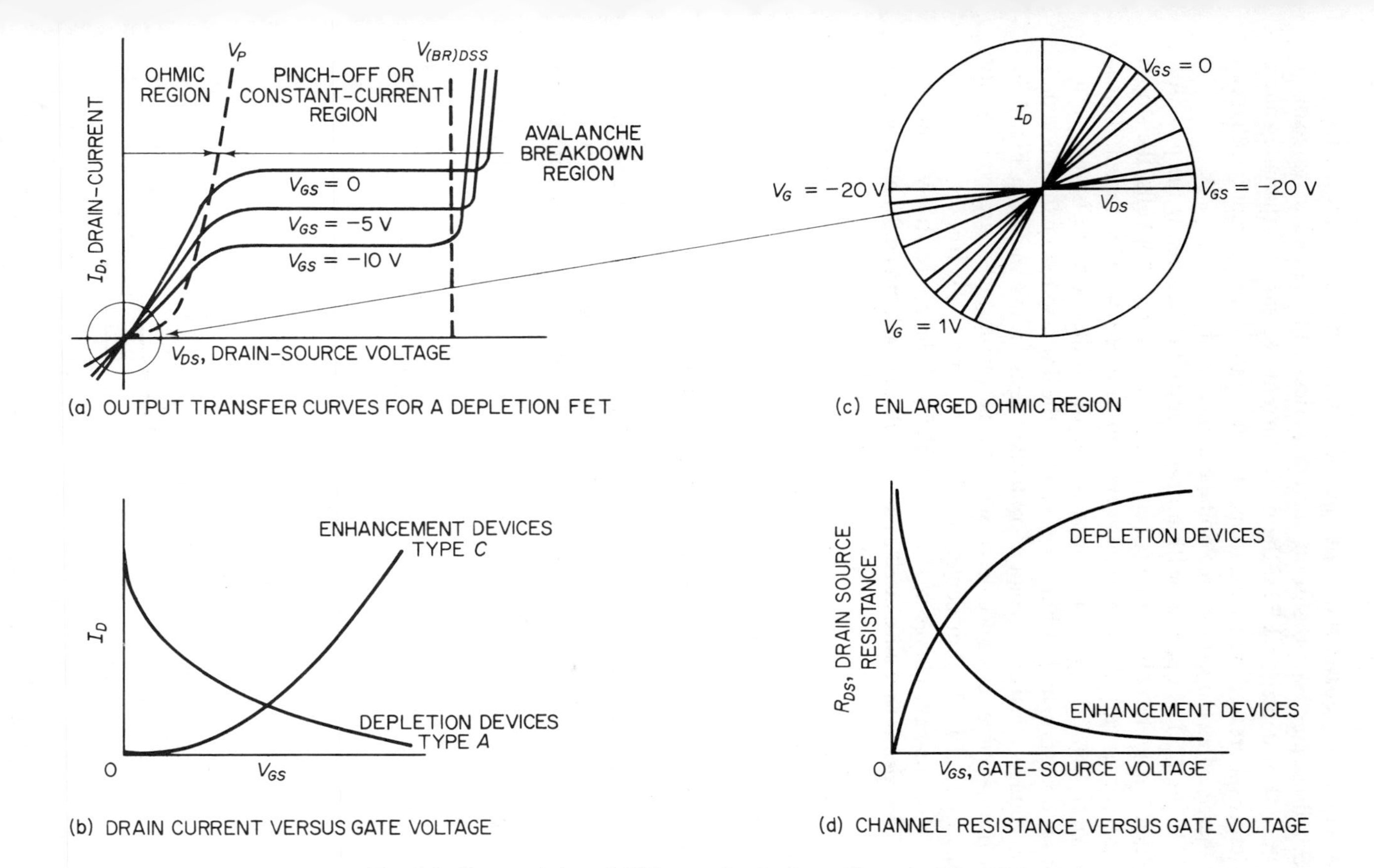

Fig. 2-3 Characteristics of FET operating regions (Courtesy Motorola)

The I_D-V_{GS} curve, shown in Fig. 2-3b and found on a typical FET data sheet, illustrates how the drain current varies with changes in gate-source voltage. For depletion FETs, the drain current decreases as the gate-source voltage is increased. For enhancement devices, the drain current is enhanced or increased as the gate-source voltage is increased.

If the FET is operated with a drain-source voltage below the pinch-off voltage, or preferably a voltage below several hundred millivolts, the slopes of the curves vary considerably as the gate-source voltage is varied. This is shown in Fig. 2-3c. Since the slope varies, the drain-source resistance varies. This is operation in the ohmic region. Here the drain-source channel is actually a voltage-variable or voltage-controlled resistor. As shown in Fig. 2-3d, the drain-source resistance decreases with increasing gate-source voltage, for enhancement FETs. For depletion FETs, the converse is true.

Note that the curves near the origin (Fig. 2-3c) are relatively symmetrical. This means that a-c as well as d-c signals can be handled. In other words, the drain-source channel is bilateral, not unilateral.

Note that an FET is generally operated in the pinch-off region for linear devices, whereas the ohmic region is used only for voltage-variable resistor applications.

2-4 Zero-Temperature-Coefficient Point

An important characteristic of all FETs is their ability to operate at a zero-temperature-coefficient point. This means that if the gate-source is biased at a specific voltage, and is held constant, the drain current will not vary with changes in temperature.

The I_D-V_{GS} curves of Fig. 2-4 show that the various curves at different

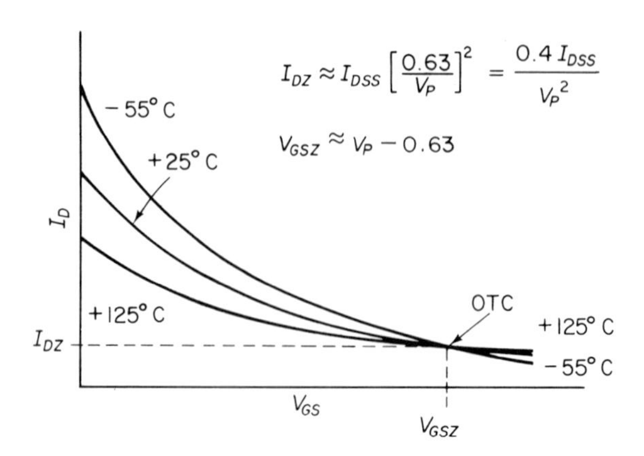

Fig. 2-4 Zero temperature coefficient of FETs (Courtesy of Motorola Inc., Semiconductor Products Division)

temperatures intersect at a common point. If the FET is operated at this value of I_D and V_{GS} (shown as I_{DZ} and V_{GSZ}), zero-temperature-coefficient operation will result.

This point varies from one FET to another and is dependent upon I_{DSS}, the zero-gate-voltage drain current, and V_P. The equations shown in Fig. 2-4 provide good approximations of the zero-temperature-coefficient point. For example, if the pinch-off voltage V_P is 1 V, the zero-temperature-coefficient mode will be obtained if the gate-source voltage V_{GS} is 0.37 V (1 − 0.63 = 0.37).

Typically, junction FETs (generally referred to as JFETs) show the zero-temperature-coefficient (0TC) characteristic over a wide range of temperature, approximately 150°C. Insulated-gate FETS (referred to as IGFETs or MOSFETs) are limited to a much narrower range, approximately 50°C.

It is sometimes assumed that the forward transadmittance (Y_{fs} or Y_{21}, Refer to Sec. 2-6) of the FET does not vary with temperature, particularly if the FET is biased at the 0TC point. However, this is not correct. The transadmittance of the FET is the slope of the I_D-V_{GS} curve. The curve of Fig. 2-4 shows that the slope varies with temperature at every point on the curve.

Figures 2-5 and 2-6 illustrate the temperature coefficients for a typical JFET.

Keep in mind that *it is not always practical to operate a FET at the zero-temperature-coefficient point*. For example, assume that the required V_{GS} to produce 0TC is 0.37 V, and the FET is to operate as an amplifier with 0.5-V input signals. A part of the input signal will be clipped. Or, assume that the circuit is to be self-biased with a source resistor (Sec. 2-5). An increase in bias resistance to produce 0TC could reduce gain.

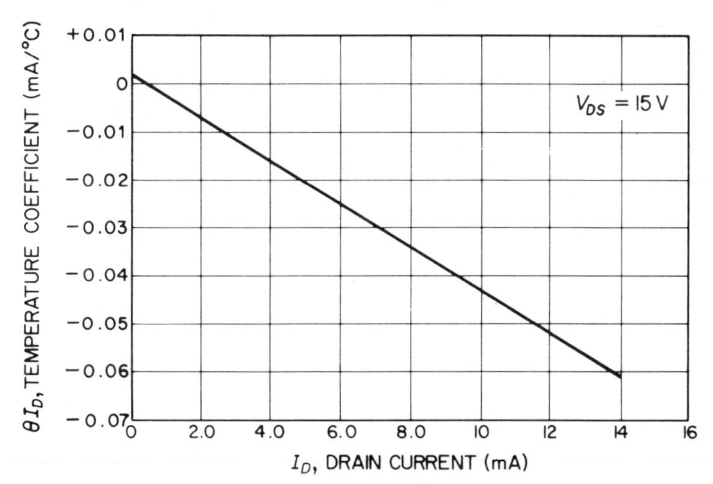

Fig. 2-5 Drain current temperature coefficient versus drain current (Courtesy of Motorola Inc., Semiconductor Products Division)

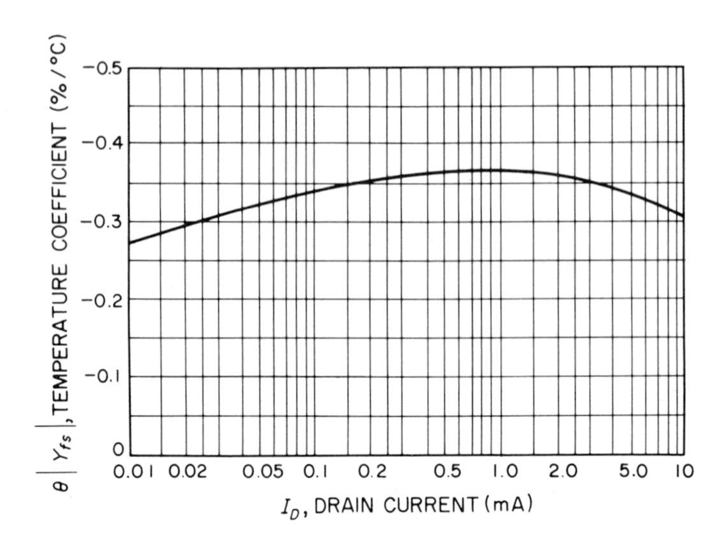

Fig. 2-6 Forward transadmittance temperature coefficient versus drain current (Courtesy of Motorola Inc., Semiconductor Products Division)

2-4-1 Practical Methods for Finding 0TC

The values of I_D and V_{GS} that produce 0TC can be found by using data-sheet curves, or by equations, as shown in Fig. 2-4; however, these values are typical approximations.

A more practical method for determining I_{DZ} requires a soldering tool, a coolant (a can of Freon), and a curve tracer (such as the Tektronix Type 575). By placing a 1000-ohm resistor across the base and emitter terminals of the curve tracer test socket, the constant-current base drive is converted to a relatively constant voltage for driving the FET gate. The curve tracer is adjusted to display the I_D-V_{DS} output family (Fig. 2-3a). By alternately bringing the soldering tool near the FET, and spraying the FET with Freon, the *voltage step of V_{GS} which remains motionless* on the curve tracer can be observed. The I_D at this voltage step is I_{DZ}.

Typically, FETs with an I_{DSS} of about 10 to 20 mA will have an I_{DZ} of about 0.5 mA. Usually, I_{DZ} increases as I_{DSS} increases (but not always, and not in proportion). For example, the I_{DZ} of 300 mA FETs often shows an I_{DZ} as low as 1 mA.

2-5 FET Bias Methods

In linear circuit applications, the FET is biased by an external supply, by self-bias, or by a combination of these two techniques. This applies to all FETs, whether biased at the zero-temperature-coefficient point, or at some

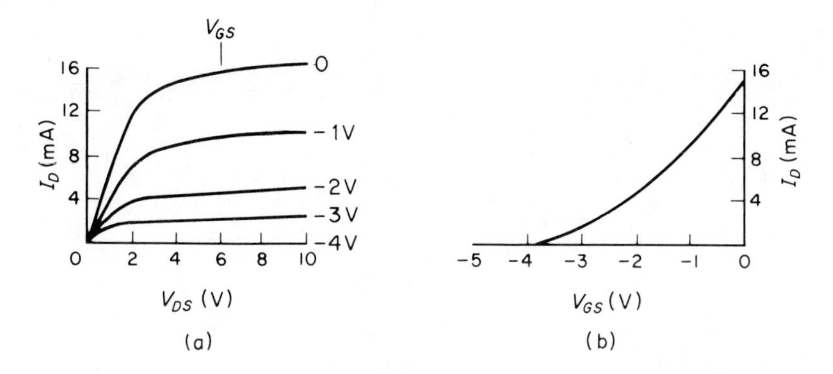

Fig. 2-7 FET characteristic curves: (a) common-source, drain characteristics; (b) transfer characteristics (Courtesy, Texas Instruments Incorporated)

other operating point. Figure 2-7a shows the familiar common-source, drain characteristic curves of a JFET (as they might appear on a typical curve tracer). For a constant level of drain-source voltage V_{DS}, drain current I_D can be plotted versus gate-source voltage V_{GS}, as shown in Fig. 2-7b. This latter curve is generally referred to as transfer characteristics. However, from a practical standpoint, either curve shows the amount of current that flows through the FET for a given gate-source voltage. For example, either curve shows that if a − 1-V bias is applied between gate and source, approximately 10 mA will flow. If the supply voltage (drain-source voltage) is 10 V, and a 500-ohm resistor is connected between the drain and supply, there is a 5-V drop across the resistor. Of course, this reduces the drain-source voltage down to 5 V, and possibly changes the characteristics. (In the example of Fig. 2-7a, there would be very little change in characteristics.)

The following paragraphs provide basic or theoretical methods by which similar curves can be used to find the correct value of bias for a given quiescent (no-signal) operating point of a FET. In Sec. 2-8 we shall discuss step-by-step procedures for finding bias values by using actual data-sheet information.

2-5-1 External Bias

Figure 2-8 shows the FET biased by an external voltage source. The input portion of this circuit is redrawn in Fig. 2-8b so that a graphical analysis may be used to determine the quiescent drain current. The graphical analysis consists of plotting the *V-I* characteristics looking into the source terminal, and the *V-I* characteristics looking into the supply-voltage terminal. When the source terminal is connected to V_0, currents I_1 and I_2 are equal. Consequently, the quiescent level of source (and drain) current is determined by the point of intersection of the *V-I* plots. For example, approximately 9 mA of current (I_{D1}) flows when V_0 (now the gate source voltage) is 1.5 V. Figure 2-8c shows this graphical analysis.

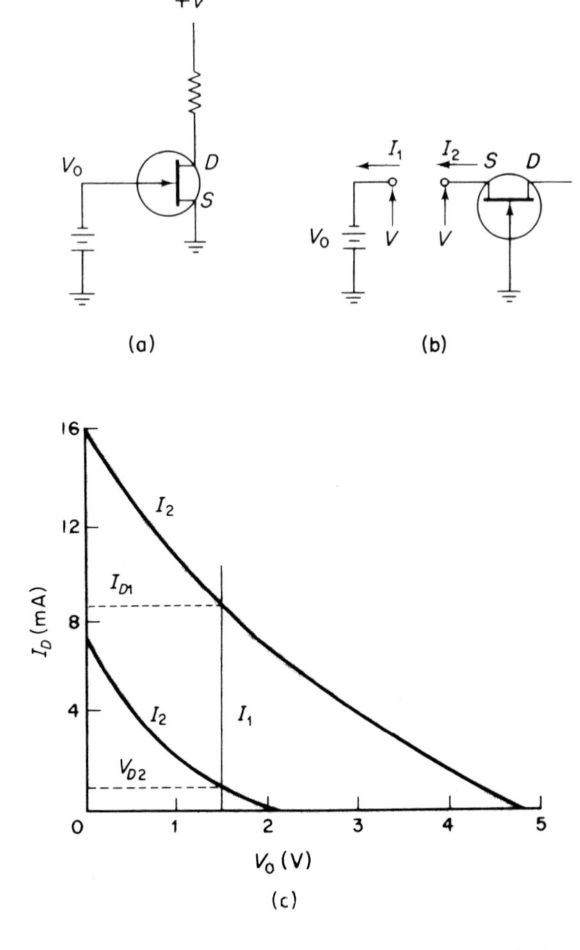

Fig. 2-8 (a) FET biased by external supply voltage; (b) equivalent circuit for bias network; (c) graphical analysis (Courtesy Texas Instruments)

Note that two I_2 curves are given. These two curves illustrate a typical spread of transfer characteristics among FETs of the same family or type. For example, with the same V_0 of 1.5 V, the lower I_2 curve shows that the current is approximately 1 mA (I_{D2}). Thus, if it is desirable that I_D be maintained at some level, a form of self-bias must be used.

2-5-2 Self-Bias

Self-bias of a FET circuit will reduce (but not eliminate) variation in quiescent levels of I_D. Figure 2-9a shows the use of a source resistor R_S to

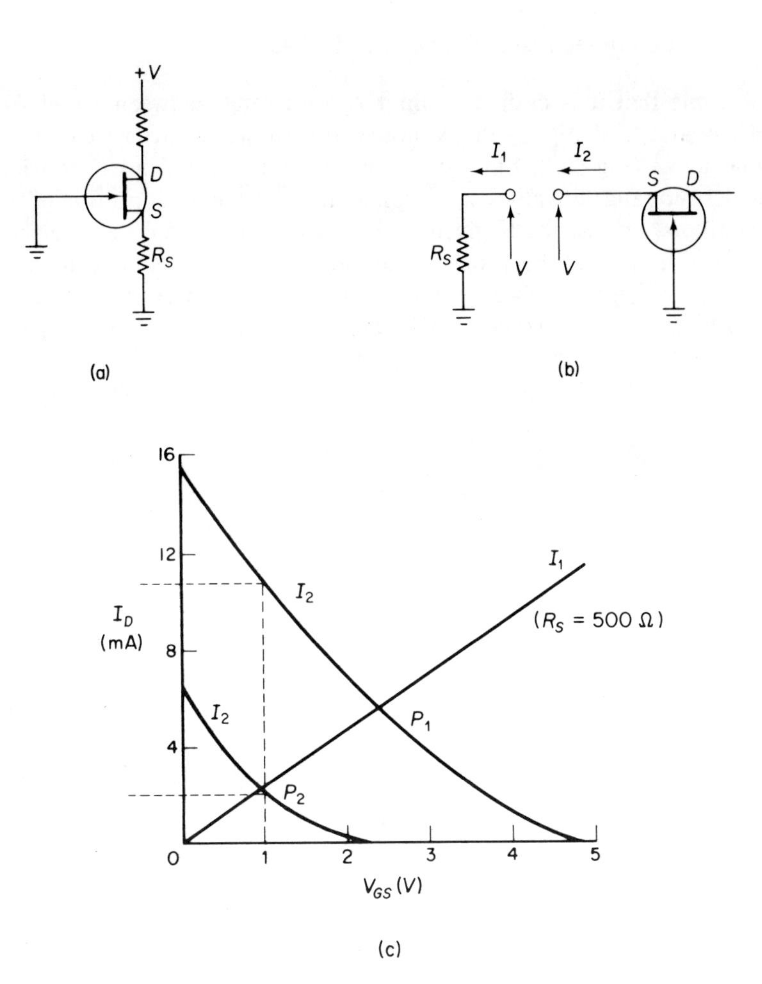

Fig. 2-9 (a) Self-bias for FET; (b) equivalent circuit for bias network; (c) graphical analysis (Courtesy, Texas Instruments Incorporated)

develop a gate-source reverse bias voltage. As I_D increases, V_{GS} becomes more negative, thus tending to prevent an increase in I_D. The input portion of the self-bias circuit is redrawn in Fig. 2-9b, and is analyzed graphically in Fig. 2-9c. The V-I characteristics for the resistor form a straight line, having a slope equal to the reciprocal of R_S, or $1/R_S$. This line intersects the two I_2 plots at points P_1 and P_2. Quiescent levels of I_D are somewhat closer for the circuit of Fig. 2-9 than for the circuit of Fig. 2-8. However, it is still possible to have a wide variation in I_D. For example, I_D could vary from about 2.5 mA to 11 mA with 1 V of V_{GS}. Thus, if it is quite essential that I_D be maintained within narrow limits, both fixed-bias and self-bias must be used.

2-5-3 Combined Fixed-Bias and Self-Bias

Assume that it is desired to limit I_D to a range between 3 and 7 mA (points A and B of Fig. 2-10). Although this cannot be accomplished with self-bias alone (Fig. 2-9), Fig. 2-10 shows two circuits which will restrict I_D to the desired range of values. A representation for the input portion of these two combination circuits is given in Fig. 2-10b. The graphical analysis in Fig. 2-10c shows that, by proper selection of power supply and resistance values, I_D can be bounded by points A and B. The step-by-step procedures for finding the values are described in Sec. 2-8.

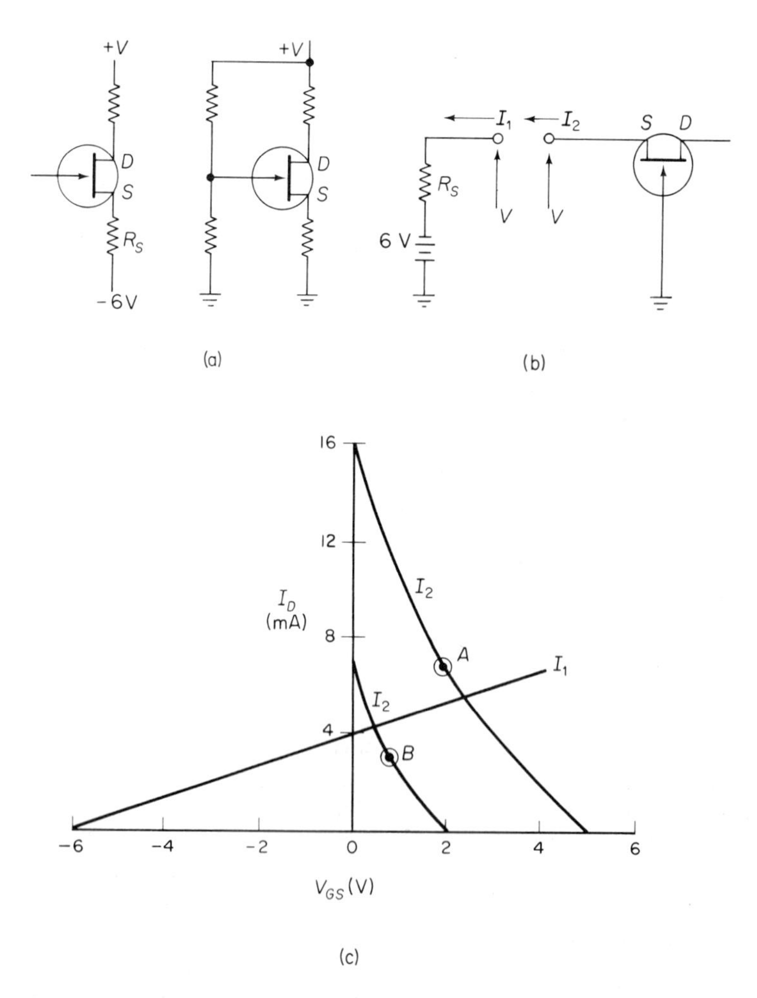

Fig. 2-10 (a) Circuits incorporating a combination of fixed- and self-bias; (b) equivalent circuit for either bias network; (c) graphical analysis (Courtesy, Texas Instruments Incorporated)

2-5-4 Constant-Current Bias

Figure 2-11a shows a bias network for a differential amplifier. Transistor Q_3 is biased as a constant-current generator in order to improve the common-mode rejection ratio of the differential amplifier (Q_1-Q_2). Figure 2-11b shows the bias network, and Fig. 2-11c shows a graphical analysis of the bias circuit. Note that the two transfer curves are shifted to the left by the amount of negative voltage appearing at the gate terminal. For example, the top

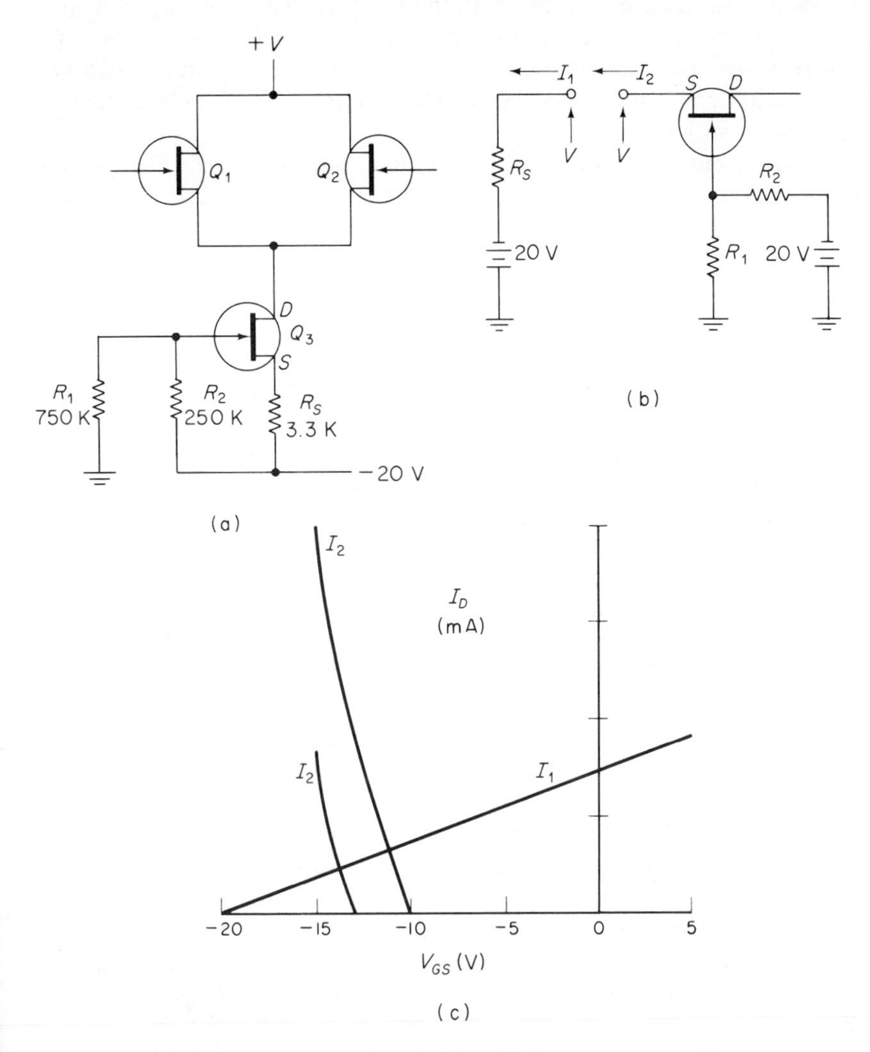

Fig. 2-11 (a) Bias circuit for differential amplifier; (b) equivalent circuit for bias network to Q_3; (c) graphical analysis (Courtesy, Texas Instruments Incorporated)

curve of Fig. 2-10 shows a maximum gate voltage of about 5 V. This same curve is shifted 15 V to the left in Fig. 2-11, and intersects the voltage axis at about −10 V. The −15-V voltage is developed across resistors R_1 and R_2 in Fig. 2-11.

2-6 Interpreting FET Data Sheets

Most of the basic design information for a particular FET can be obtained from the data sheet. There are some exceptions to this rule. For extreme high-frequency work, and in digital work where switching characteristics are of particular importance, it may be necessary to test the FET under simulated operating conditions.

In any event, it is always necessary to interpret data sheets. Each manufacturer has its own system of data sheets. It would be impractical to discuss all data sheet formats here. Instead, we shall discuss the Motorola Designers Data Sheets, and see how this information affects simplified design.

2-6-1 Data Sheet Format

Figure 2-12 (the first page of the data sheet) gives the maximum ratings of the FET, mechanical dimensions, and pin layout. Obviously, if the maximum ratings are exceeded, abnormal circuit operation will occur and the FET may be destroyed. For example, assume that 70 V is applied to the FET which has a maximum drain-source rating of 40 V. Damage, if not total destruction, will surely result. This brings up three important rules regarding maximum ratings:

Never design any circuit where the FET element (drain-source gate) is connected to a source higher than the maximum voltage rating, even through a resistance.

Always allow for some variation in supply voltage when an electronic power supply is used.

Always consider any input signal that may be applied to the FET element, in addition to the normal operating voltage.

Figure 2-13 (the second page of the data sheet) presents the electrical characteristics and operating conditions under which they were measured. Page two shows both ON and OFF characteristics, as well as "small signal characteristics." Keep in mind that all of the characteristics listed in any data sheet are based on a set of *fixed operating conditions.* If the conditions change (as they must in any practical circuit), the characteristics will change. Therefore:

Use all data-sheet characteristics as a starting point for simplified design, not as hard and fast design rules.

MOTOROLA

DESIGNERS Data Sheet

2N5358 thru 2N5364

SILICON N-CHANNEL
JUNCTION FIELD-EFFECT TRANSISTORS

. . . depletion mode devices designed primarily for general-purpose amplifier applications.

- Tightly Specified I_{DSS} Ranges –
  2:1 for All Types
- Low Noise Figure –
  NF = 1.5 dB (Typ) @ f = 100 Hz
- Complement to 2N5265 thru 2N5270
- New Designers Data Sheet with Min/Max Curves for Ease in Design

LIMIT DATA FOR "WORST CASE" DESIGNS

The Designers† Data Sheet permits the design of most circuits entirely from the information presented. Limit curves – representing boundaries for device characteristics – are given to facilitate "worst case" design.

MAXIMUM RATINGS

Rating	Symbol	Value	Unit
Forward Gate Current	$I_{G(f)}$	10	mAdc
Reverse Gate-Source Voltage	$V_{GS(r)}$	40	Vdc
Drain-Gate Voltage	V_{DG}	40	Vdc
Total Device Dissipation @ T_A = 25°C Derate above 25°C	P_D	300 2.0	mW mW/°C
Storage Temperature Range	T_{stg}	-65 to +200	°C
Operating Junction Temperature Range	T_J	-65 to +175	°C

†Trademark of Motorola Inc.

N-CHANNEL
JUNCTION FIELD-EFFECT
TRANSISTORS

DS 5278

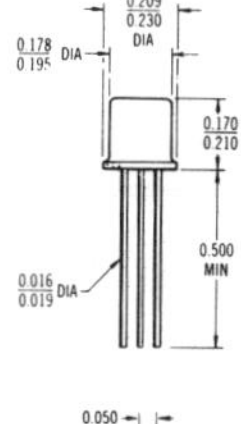

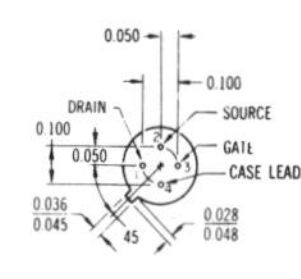

TO-72
CASE 20 (3)

Si FIELD-EFFECT TRANSISTORS
2N5358 thru 2N5364
DS 5278

Fig. 2-12 First page of Motorola Designers Data Sheets (Courtesy of Motorola Inc., Semiconductor Products Division)

ELECTRICAL CHARACTERISTICS (T_A = 25°C unless otherwise noted)

Characteristic		Symbol	Min	Max	Unit
OFF CHARACTERISTICS					
Gate-Source Breakdown Voltage (I_G = 10 μAdc, V_{DS} = 0)		$V_{(BR)GSS}$	40	-	Vdc
Gate-Source Cutoff Voltage (V_{DS} = 15 Vdc, I_D = 100 nAdc)	2N5358	$V_{GS(off)}$	0.5	3.0	Vdc
	2N5359		0.8	4.0	
	2N5360		0.8	4.0	
	2N5361		1.0	6.0	
	2N5362		2.0	7.0	
	2N5363		2.5	8.0	
	2N5364		2.5	8.0	
Gate Reverse Current (V_{GS} = 20 Vdc, V_{DS} = 0)		I_{GSS}	-	0.1	nAdc
(V_{GS} = 20 Vdc, V_{DS} = 0, T_A = 150°C)			-	0.1	μAdc
ON CHARACTERISTICS					
Zero-Gate Voltage Drain Current (V_{DS} = 15 Vdc, V_{GS} = 0)	2N5358	I_{DSS}	0.5	1.0	mAdc
	2N5359		0.8	1.6	
	2N5360		1.5	3.0	
	2N5361		2.5	5.0	
	2N5362		4.0	8.0	
	2N5363		7.0	14	
	2N5364		9.0	18	
Gate-Source Voltage (V_{DS} = 15 Vdc, I_D = 50 μAdc)	2N5358	V_{GS}	0.3	1.5	Vdc
(V_{DS} = 15 Vdc, I_D = 80 μAdc)	2N5359		0.4	2.0	
(V_{DS} = 15 Vdc, I_D = 150 μAdc)	2N5360		0.5	2.5	
(V_{DS} = 15 Vdc, I_D = 250 μAdc)	2N5361		1.0	5.0	
(V_{DS} = 15 Vdc, I_D = 400 μAdc)	2N5362		1.3	5.0	
(V_{DS} = 15 Vdc, I_D = 700 μAdc)	2N5363		2.0	6.0	
(V_{DS} = 15 Vdc, I_D = 900 μAdc)	2N5364		2.0	6.0	
SMALL-SIGNAL CHARACTERISTICS					
Forward Transadmittance (V_{DS} = 15 Vdc, V_{GS} = 0, f = 1.0 kHz)	2N5358	$\|y_{fs}\|$	1000	3000	μmhos
	2N5359		1200	3600	
	2N5360		1400	4200	
	2N5361		1500	4500	
	2N5362		2000	5500	
	2N5363		2500	6000	
	2N5364		2700	6500	
Forward Transconductance (V_{DS} = 15 Vdc, V_{GS} = 0 Vdc, f = 100 MHz)	2N5358	$Re(y_{fs})$	800	-	μmhos
	2N5359		900	-	
	2N5360		1400	-	
	2N5361		1700	-	
	2N5362		1900	-	
	2N5363		2100	-	
	2N5364		2200	-	
Output Admittance (V_{DS} = 15 Vdc, V_{GS} = 0, f = 1.0 kHz)	2N5358, 2N5359	$\|y_{os}\|$	-	10	μmhos
	2N5360, 2N5361			20	
	2N5362, 2N5363			40	
	2N5364			60	
Input Capacitance (V_{DS} = 15 Vdc, V_{GS} = 0, f = 1.0 MHz)		C_{iss}	-	6.0	pF
Reverse Transfer Capacitance (V_{DS} = 15 Vdc, V_{GS} = 0, f = 1.0 MHz)		C_{rss}	-	2.0	pF
Common-Source Noise Figure (V_{DS} = 15 Vdc, V_{GS} = 0, R_G = 1.0 Megohm, f = 100 Hz, BW = 1.0 Hz)		NF	-	2.5	dB
Equivalent Short-Circuit Input Noise Voltage (V_{DS} = 15 Vdc, V_{GS} = 0, f = 100 Hz, BW = 1.0 Hz)		e_n	-	115	$nV/\sqrt{Hz}$

MOTOROLA Semiconductor Products Inc. 

Fig. 2-13 Second page of Motorola Designers Data Sheets (Courtesy of Motorola Inc., Semiconductor Products Division)

2-6-2 FET *Characteristics*

Forward gate current $I_{G(f)}$ is the maximum recommended forward current through the gate terminal. This is a limiting factor in some applications, and is caused by a large forward bias current on the gate. When this condition occurs, the gate current must be limited, or degeneration of the FET will occur. A resistor in series with the gate will limit the current, but its value will determine the variance of gate bias as affected by the gate leakage current.

Total device dissipation P_D is the maximum power that can be dissipated within the device at 25°C without exceeding the maximum allowable internal temperature (200°C in this case). The power is derated according to the thermal resistance value of 2 mW/C°. For example, at 125°C, the power dissipated in the FET must not exceed 100 mW (125°C − 25°C = 100; 100 × 2 = 200; 300 mW − 200 mW = 100 mW). Operation above the maximum value could damage the device.

Gate-source breakdown voltage $V_{(BR)GSS}$ is the breakdown voltage from gate to source with the drain and source shorted. With drain and source shorted, the gate-channel junction also meets the breakdown specification since the drain and source are the connections to the channel. For the designer, this means that the *drain and source may be interchanged, for symmetrical devices*, without fear of individual junction breakdown. The minimum $V_{(BR)GSS}$ voltage for the 2N5358 through 2N5364 FETs shown in Fig. 2-13 is 40 V. The relatively high value is useful where large signal swings are required, high voltage power supplies are used, or high voltage transients are encountered.

Gate-source cutoff voltage $V_{GS(OFF)}$ is defined as the gate-to-source voltage required to reduce the drain current to 0.01 or preferably 0.001 of the minimum I_{DSS} value. A standard I_D of 1 nA is used for the 2N5358 through 2N5364.

Pinch-off voltage V_P is essentially the same as $V_{GS(OFF)}$, only measured in a different manner. V_P is shown in Fig. 2-14, and is the drain-to-source voltage at which the drain current increases very little for an increase in drain-to-source voltage at $V_{GS} = 0$. Most of the equations used to describe the operation of a FET use V_P, but the value of $V_{GS(OFF)}$ can be used instead.

An *approximate value* for $V_{GS(OFF)}$ can also be determined from the transfer characteristic curve shown in Fig. 2-15. $V_{GS(OFF)}$ is found by taking the *slope of the curve* at $V_{GS} = 0$ (where the curve intersects the I_D axis), and extending the slope to the V_{GS} axis. Twice the value of this intercept is $V_{GS(OFF)}$. From Fig. 2-15, V_{GS} minimum is about 2 × 1.25 V, or 2.5 V. V_{GS} maximum is about 2 × 1.8 V, or 3.6 V at 25°C. This technique can be used on any of the I_D-V_{GS} curves for other temperatures.

$V_{GS(OFF)}$ must be used to determine the minimum operational drain-to source voltage. As a rule of thumb, when V_{DS} is greater than 1.5 × $V_{GS(OFF)}$, operation in the active region is assured.

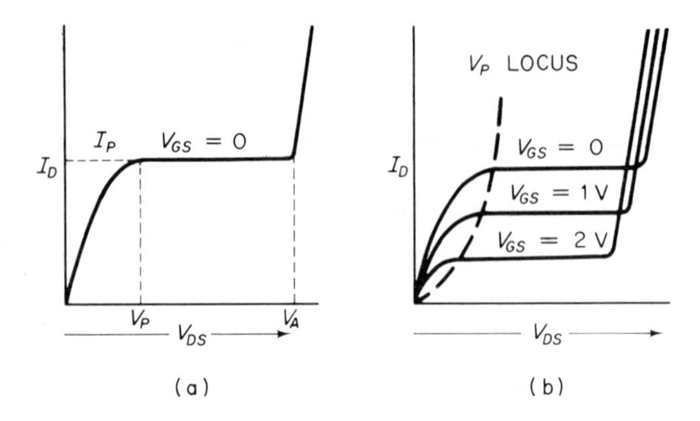

Fig. 2-14 FET output characteristics (Courtesy of Motorola Inc., Semiconductor Products Division)

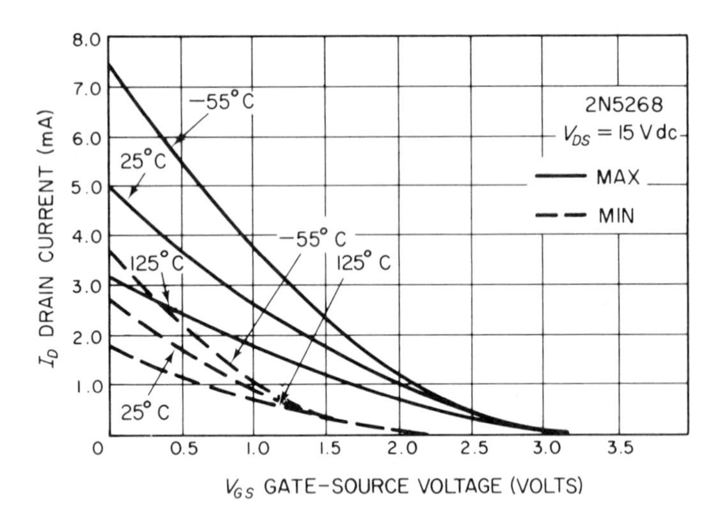

Fig. 2-15 Forward transfer characteristic curves for min/max I_{DSS} limits (Courtesy of Motorola Inc., Semiconductor Products Division)

Gate leakage (reverse) current I_{GSS} is defined as the gate-channel leakage with the drain shorted to the source, and is a measure of the static short-circuit input impedance. Since gate-to-channel is a reverse-biased diode junction (for a JFET), I_{GSS} doubles (approximately) every 15°C increase in temperature, and is proportional to the square root of the applied voltage.

For the 2N5358 through 2N5364, I_{GSS} is measured at 20 V, which is much higher than the normal gate-to-source voltage. On the other hand, a gate-drain voltage of 20 V is perhaps typical so that I_{GSS} exceeds the worst-case gate leakage current for most applications.

Zero gate voltage drain current I_{DSS} is defined as the drain-to-source current with the gate shorted to the source at a specified drain-source voltage. I_{DSS} is a basic parameter for JFETs and is considered to be a figure of merit.

For the 2N5358 through 2N5364, each FET has a min/max I_{DSS} ratio of 2 : 1. This tight ratio has numerous advantages. Especially for worst-case design (where higher stage gain can be obtained), there is less distortion and less source degeneration. With a tight ratio, a smaller source resistor achieves the desired results. With a larger I_{DSS} ratio, the $-55°C$ curve, as shown in Fig. 2-15, will shift to the right. This decreases the slope of the load line, for the same I_D, thus requiring a larger value for R_S (source resistor).

Minimum/maximum transfer characteristic curves for the 2N5268 (a P-channel complement to the 2N5358 through 2N5364 series), at three different temperatures are shown in Fig. 2-15. Using these curves, worst-case design is reduced to a simple graphical technique and requires no further testing of the FET.

Gate-source voltage V_{GS} is a range of gate-to-source voltages with 0.1 I_{DSS} drain current flowing. The specified drain-to-source voltage is the same as for I_{DSS}. This characteristic gives the designer the min/max variation in V_{GS} for different FETs for a given I_D and V_{DS}.

Forward transadmittance Y_{fs} is defined as the magnitude of the common-source forward transfer admittance. Y_{fs} shows the relationship between input signal voltage and output signal current and is a key dynamic characteristic for all FETs. Y_{fs} is specified at 1 kHz, with d-c operating conditions the same as for I_{DSS}.

At 1 kHz, Y_{fs} is almost entirely real. Thus, Y_{fs} at 1 kHz $= Y_{fs}$. At higher frequencies, Y_{fs} includes the effects of gate-to-drain capacitance and may be misleadingly high. For high-frequency operation, the real part of transadmittance $\mathrm{Re}(Y_{fs})$, as discussed in later paragraphs, should be used.

Figure 2-16 shows Y_{fs} versus I_D with typical and minimum value shown.

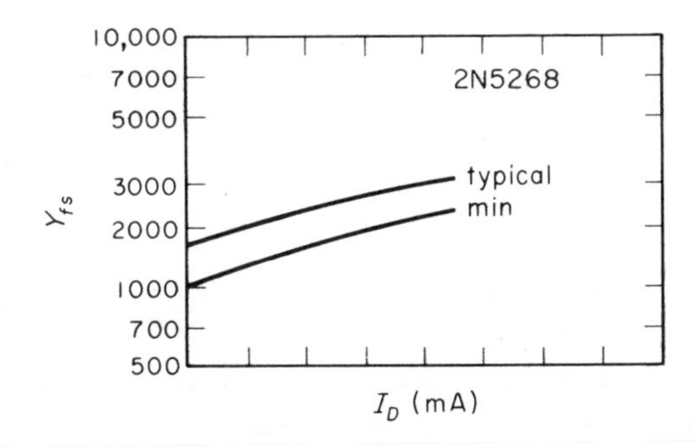

Fig. 2-16 Typical and minimum forward admittance Y_{fs} (Courtesy of Motorola Inc., Semiconductor Products Division)

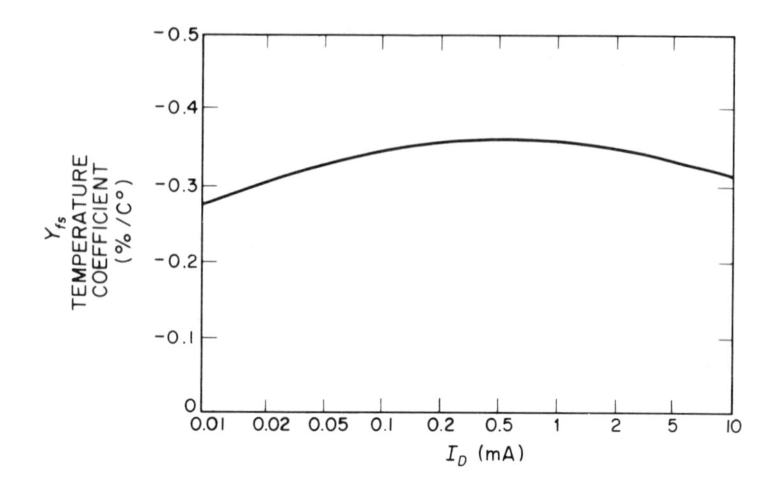

Fig. 2-17 Forward transadmittance coefficient versus drain current (Courtesy of Motorola Inc., Semiconductor Products Division)

The curves end sharply where $I_D = I_{DSS}$ and, at this value, Y_{fs} is at its maximum.

Figure 2-17 shows the typical temperature coefficient of Y_{fs} versus drain current. For this curve $V_{DS} = 15$ V, and V_{GS} is varied to obtain the variations of I_D.

Forward transconductance Re(Y_{fs}) is defined as the common-source forward transfer conductance (drain current versus gate voltage). For high-frequency applications, Re(Y_{fs}) is considered a figure of merit. The d-c operating conditions are the same as for Y_{fs}, but the test frequency is 100 mHz. All other factors being equal, an increase in Re(Y_{fs}) produces an increase in the voltage gain produced by a FET amplifier stage.

In comparing Re(Y_{fs}) with Y_{fs}, the minimum values of the two are quite close, considering the difference in frequency at which the measurements are made. At high frequencies, about 30 MHz and above, Y_{fs} will increase due to the effect of gate-drain capacitance C_{gd} so that Y_{fs} will be misleadingly high.

Output admittance Y_{os} is defined as the magnitude of the common-source output admittance and is measured at the same operating conditions and frequency as Y_{fs}. Since Y_{os} is a complex number at low frequencies, only the magnitude is specified. For higher frequencies, Y_{os} can be calculated by using r_{oss}, as shown in Fig. 2-18.

The common-source output resistance r_{oss} is the real part of Y_{os}. Figure 2-19 shows the variation of r_{oss} versus I_D for several values of I_{DSS}. These are also measured at 1 kHz, $V_{DS} = 15$ V, and V_{GS} varied to obtain the different values of I_D. Note that the lower I_{DSS} units have a higher output resistance for the same drain current.

Input capacitance C_{iss} is the common-source input capacitance with the

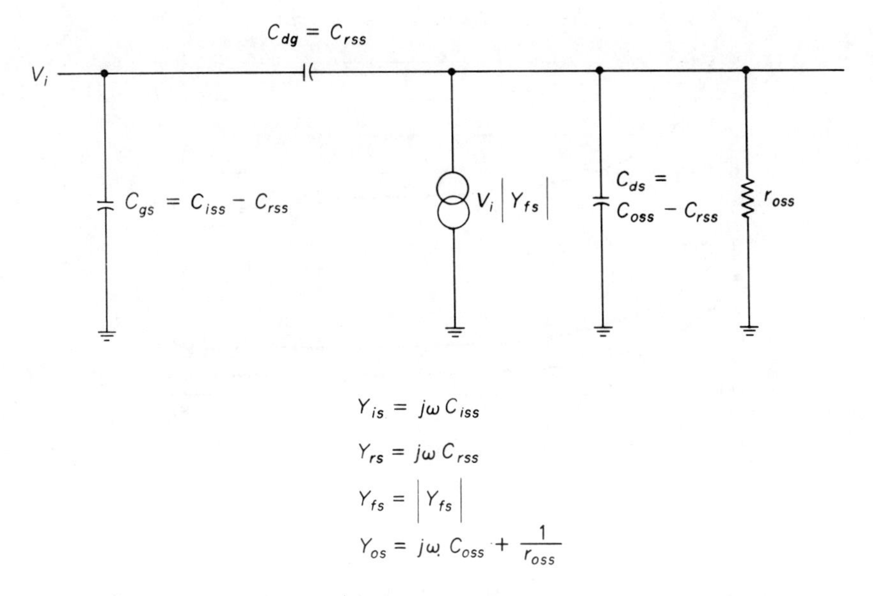

$$Y_{is} = j\omega C_{iss}$$
$$Y_{rs} = j\omega C_{rss}$$
$$Y_{fs} = |Y_{fs}|$$
$$Y_{os} = j\omega C_{oss} + \frac{1}{r_{oss}}$$

Fig. 2-18 Equivalent low-frequency circuit (Courtesy of Motorola Inc., Semiconductor Products Division)

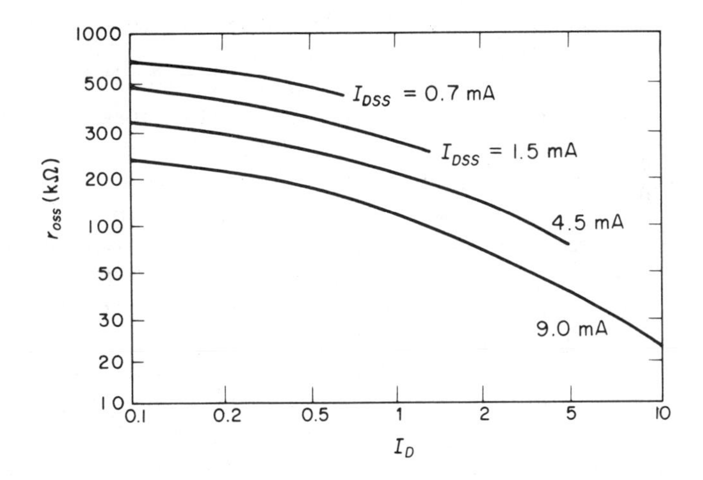

Fig. 2-19 Output resistance versus drain current (Courtesy of Motorola Inc., Semiconductor Products Division)

output shorted and is used in place of Y_{iss}, the short-circuit input admittance. Y_{iss} is entirely capacitive at low frequencies, since the input is a reverse-biased silicon diode. The real part of Y_{iss} could be calculated from I_{GSS} but is negligible, even at low frequencies.

Reverse transfer capacitance C_{rss} is defined as the common source reverse transfer capacitance with the input shorted. C_{rss} is used in place of

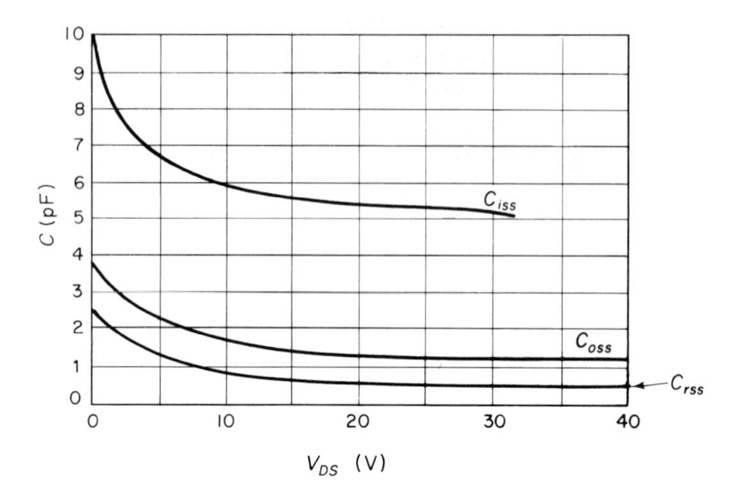

Fig. 2-20 Capacitance versus drain-source voltage (Courtesy of Motorola Inc., Semiconductor Products Division)

Y_{rs}, the short-circuit reverse transfer admittance, since Y_{rs} is almost entirely capacitive over the useful frequency range of the FET. Figure 2-20 shows C_{iss}, C_{rss}, and C_{oss} as functions of V_{DS}.

A notable feature of the Motorola Designers Data Sheet is the equivalent circuit of the FET, as shown in Fig. 2-18. For the 2N5358 through 2N-5364, the circuit is valid up to about 30 MHz and shows how to obtain all of the short-circuit Y parameters from the data given in the data sheet. Although all of this information may not be necessary for simple linear FET amplifiers used at audio frequencies, the data can be of great value for r-f amplifiers, oscillators, and the like, as discussed in later sections of this chapter.

Common-source noise figure NF represents a ratio between input signal-to-noise and output signal-to-noise and is measured in dB. Short-circuit input noise voltage e_n is the equivalent short-circuit input noise expressed in volts per root cycle.

NF, as specified, includes the effects of e_n and i_n, where i_n is the equivalent open-circuit input noise current. For a FET, the contribution of i_n is small compared to e_n. As shown in Fig. 2-21, NF attains its highest value for a small generator resistance and decreases for increasing generator resistance, indicating a large noise contribution from the noise-voltage generator. For this reason, NF and e_n are specified, and i_n is neglected. NF is independent of operating current and proportional to voltage. However, the voltage effects are slight over the normal operating range.

Figure 2-22 is a nomograph for converting noise figure to equivalent input noise voltage for different generator source impedances R_S. This nomograph can be used with any FET. Since NF and e_n are frequency dependent, Fig. 2-22 must be used in conjunction with Fig. 2-23 (noise figure versus

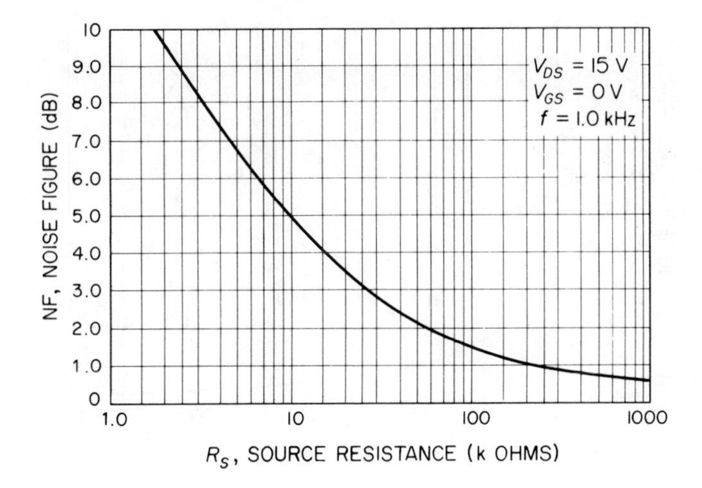

Fig. 2-21 Noise figure versus source resistance (Courtesy of Motorola Inc., Semiconductor Products Division)

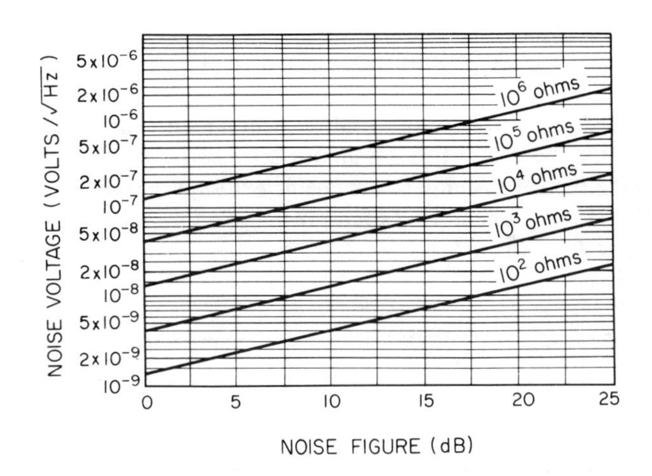

Fig. 2-22 Noise figure conversion chart (Courtesy of Motorola Inc., Semiconductor Products Division)

frequency at a specified source impedance) to determine e_n. For example, to find the input noise at 50 Hz, with an R_S of 1 megohm, the NF (from Fig. 2-23) is about 1.5 dB. Next, from Fig. 2-22, e_n is found to be 1.5×10^{-7} (on the 10^6 ohms curve) volts, or about 150 nV. If the stage has a voltage gain of 10, the output noise is 10×150 nV or 1.5 μV of 50-Hz noise.

Figure 2-21 shows NF versus generator resistance at a fixed frequency. The point common to Figs. 2-23 and 2-21 is at $R_S = 1$ megohm and $f = 1$ kHz. For other values of R_S and f, NF can be estimated by using both figures; then e_n is found, as before, from Fig. 2-22.

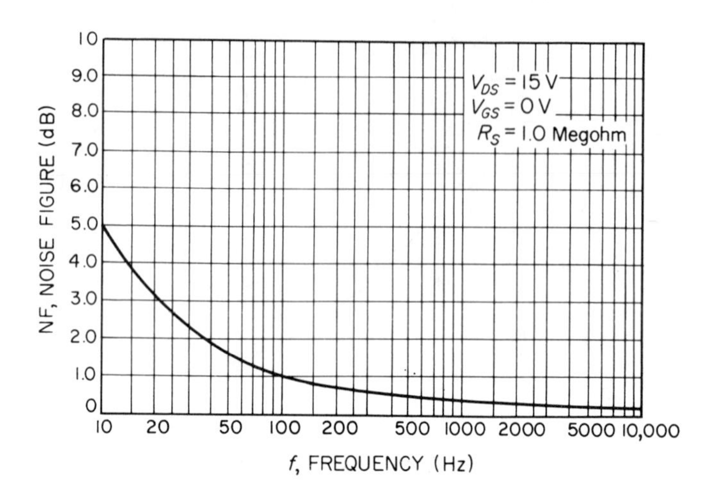

Fig. 2-23 Noise figure versus frequency (Courtesy of Motorola Inc., Semiconductor Products Division)

2-7 Temperature Coefficient of I_D or V_{GS}

While there is no single specification dealing with the zero-tempeature-coefficient (0TC) effect of FETs, its importance warrants discussion. Figure 2-24 shows the typical I_D temperature coefficient θI_D versus I_D, for the 2N5265 through 2N5270. All of the FETs have a temperature coefficient of zero at approximately $I_D = 0.5$ mA.

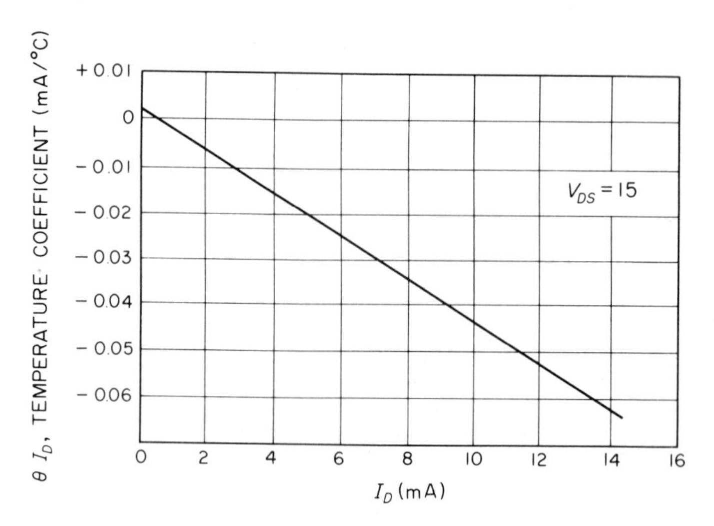

Fig. 2-24 Drain current temperature coefficient versus drain current (Courtesy of Motorola Inc., Semiconductor Products Division)

The positive temperaure coefficient occurring at small values of drain current is caused by a change in the width of the thermally generated depletion layer at the gate-channel junction, and changes at a rate equivalent to −2.2 mV/°C across the gate-source junction. This is the same value and phenomenon as the forward-biased base-emitter junction of a bipolar transistor. The negative temperature occurs at high values of I_D due to a decrease in carrier mobility (increasing resistivity) in the channel. At some value of I_D these two effects cancel each other, and zero TC of I_D or V_{GS} occurs. (Refer to Sec. 2-4.)

The design equations for 0TC operation are:

$$V_{GSZ} \approx V_P - 0.63 \tag{2-1}$$

$$I_{DZ} \approx I_{DSS}\left(\frac{0.63}{V_P}\right)^2 \tag{2-2}$$

$$\text{Drift in mV/°C} \approx 2.2\left(1 \frac{\sqrt{I_D}}{I_{DZ}}\right) \tag{2-3}$$

where subscript Z indicates the 0TC point for the parameter.

Equation (2-3) is valid near, but not at, I_{DZ}. Equations (2-1) and (2-2) must both be satisfied for 0TC operation. Since V_P and I_{DSS} vary from device to device, an adjustment for V_{GS} or I_D is necessary. Note that resistor temperature coefficients and the effects of I_{GSS} with temperature must also be considered.

2-8 Designing the Basic FET Stage

The following example is presented to show how the transfer characteristic curves from the Motorola Designers Data Sheet are used in a worst-case design of a typical FET bias network. The circuit involved is shown in Fig. 2-25. Note that both fixed-bias and self-bias (Sec. 2-5) are used. As discussed in later sections, the basic bias network is modified as necessary to produce an FET stage with desired characteristics (stage gain, input/output impedance, etc.).

Keep one point in mind when studying the following bias scheme. The purpose of the basic bias circuit is to establish a given I_D, and to maintain that I_D (plus or minus some given tolerance) over a given temperature range. Generally, this is to keep the FET at the 0TC operating point. No consideration is given to operating point, gain, impedance, etc. For example, if the FET is used as a linear amplifier, the output (drain terminal) should be at one-half the supply voltage, if maximum output voltage swing is wanted. Assume that supply is 30 V and that the drain is set at 15 V under no-signal conditions. Then the output can swing almost 30 V, from 0 to 30 V. Since FETs rarely operate under static conditions, the basic bias circuit is used as a reference

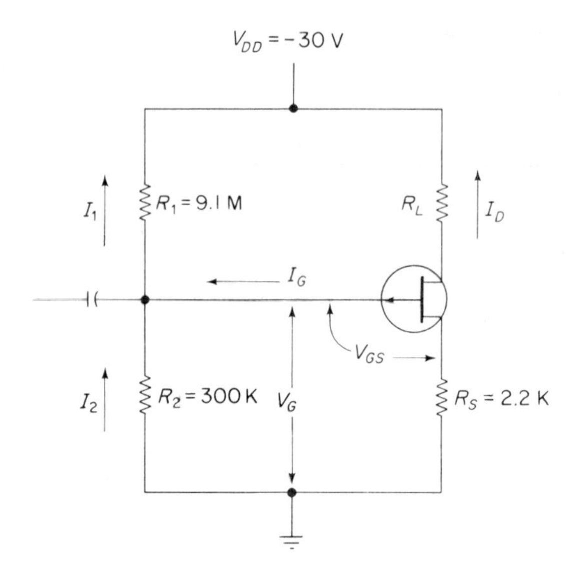

Fig. 2-25 The basic FET stage (Courtesy of Motorola Inc., Semiconductor Products Division)

or starting point for design. The actual circuit configuration, and especially the bias circuit values, should be selected on the basis of dynamic circuit conditions (desired output voltage swing, expected input signal level, etc.).

2-8-1 Design Example

Assume that the circuit of Fig. 2-25 is to maintain I_D at 1 ± 0.25 mA, over a temperature range from -55°C to $+125$°C, with a supply voltage V_{DD} of 30 V, using a 2N5269.

The first step is to draw a $1/R_S$ load line on the FET transfer characteristics, as shown in Fig. 2-26. (Note that this illustration is similar to that of Fig. 2-10.) As shown by the equations, the value of R_S is set by the limits of V_{GS} and I_D. The value of $V_{GS\ \min}$ is the point where the $I_{D\ \min}$ of 0.75 mA crosses the high-temperature limit curve of $+125$°C, or approximately 0.8 V. The value of $V_{GS\ \max}$ is the point where the $I_{D\ \max}$ of 1.25 mA crosses the low-temperature limit curve of -55°C, or approximately 1.9 V. Using these values, the first trial value for R_S is:

$$R_S = \frac{1.9 - 0.8}{1.25 - 0.75} = 2.2\ \text{k}\Omega$$

The fixed-bias voltage V_G is determined from the intercept of the $1/R_S$ load line with the V_{GS} axis, and is computed by using the same set of values as shown in the Fig. 2-26 equations:

$$V_G = \frac{0.75 \times 1.9 - 1.25 \times 0.80}{0.5} = 0.85\ \text{V}$$

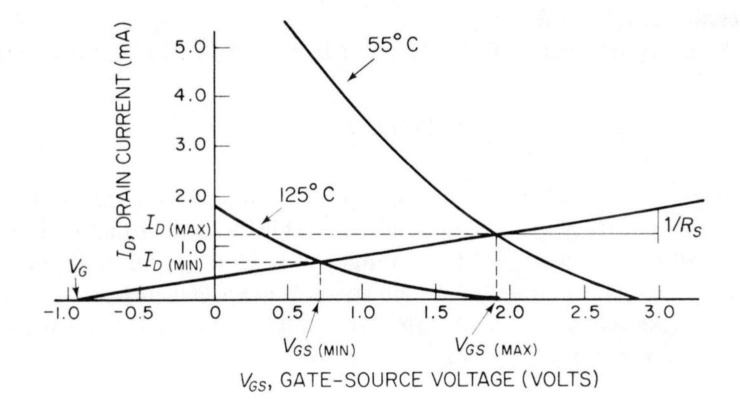

$$R_S = \frac{V_{GS\,(MAX)} - V_{GS\,(MIN)}}{I_{D\,(MAX)} - I_{D\,(MIN)}}$$

$$V_{R_L} = V_{DD} - (1.5 \times V_{GS\,(OFF)}) - (I_{DC\,(MAX)} \times R_S)$$

$$R_2 = \frac{V_G \cdot R_1}{V_{DD} - V_G} \qquad R_{L\,(MAX)} = \frac{V_{R_L}}{I_{DC\,(MAX)}}$$

$$V_G = \frac{I_{D\,(MIN)} \cdot V_{GS\,(MAX)} - I_{D\,(MAX)} V_{GS\,(MIN)}}{I_{D\,(MAX)} - I_{D\,(MIN)}}$$

Fig. 2-26 $1/R_S$ load line for basic FET stage (Courtesy of Motorola Inc., Semiconductor Products Division)

The maximum value of R_1 is determined by the maximum gate reverse current. As shown on the 2N5268 data sheet, the maximum gate reverse current at 150°C and 30-V supply voltage is 2 μA. Gate reverse current variations with temperature follow the pattern of all silicon devices. As a rule, it can be assumed that it will double with each 15°C temperature rise. Thus, at 125°C the gate reverse current should be about 0.5 μA.

The variation in V_G versus temperature will not be too great if a value for R_1 is chosen such that I_1 (Fig. 2-25) is at least six times greater than the maximum reverse current. Assuming a value of 9.1 megohms for R_1, I_1 is 3.2 μA, which satisfies the above requirement. ($6 \times 0.5\ \mu A = 3\ \mu A$.)

The value of R_2 is found from a simple voltage divider relationship, ignoring the effect of I_G, as shown in the Fig. 2-26 equations:

$$R_2 = \frac{0.85 \times 9.1 \times 10^6}{30 - 0.85} \approx 300\text{K}$$

The maximum value of R_L is determined by the voltage drop across R_L and $I_{D\,max}$. As shown by the Fig. 2-26 equations, the voltage drop across R_L (or V_{RL}) is:

$$V_{RL} = 30 - (1.5 \times 3.6\text{ V}) - (1.25 \times 2.2) = 21.85\text{ V}$$

Note that $I_{D\ max}$ is shown in Fig. 2-26, and $V_{GS(off)}$ is calculated by using the technique of Sec. 2-6-2.

With V_{RL} established at 21.85 V, and an $I_{D\ max}$ of 1.25 mA, the maximum value of R_L is:

$$R_{L\ max} = \frac{21.85}{1.25} = 17.48\ \text{k}\Omega$$

Assuming a standard 1 per cent resistor, a value of 16.9 kΩ can be used.

With all of the bias resistance values established, the circuit of Fig. 2-25 can be converted to a single FET amplifier stage by the addition of input and output coupling capacitors. This conversion is described in Sec. 2-10. However, before going into a complete FET circuit, we shall discuss the small-signal characteristics of the basic FET stage.

2-9 Small-Signal Analysis of Basic FET Stage

Small-signal analysis of a FET amplifier stage is easily accomplished with reasonable accuracy by using a few simple equations. The FET model used for the analysis is shown in Fig. 2-27. This model differs from the previous model (Fig. 2-18) in that C_{rss} is omitted. By omitting all capacitance and using only the real parts of Y_{is}, Y_{fs}, and Y_{os}, the model and the accompanying equations are useful up to about 100 kHz. If capacitance effects are included, the equations are useful up to several megahertz.

Figures 2-28, 2-29, and 2-30 show schematics for common-source, common-drain (source-follower), and common-gate circuits. In addition, the *exact* and *approximate* equations for voltage gain, input impedance, and output impedance are included. (The equations for finding the resistance values of R_1, R_2, R_S, and R_L are discussed in Sec. 2-8.)

For the common-source circuit (Fig. 2-28), the omission of C_{rss} will affect the equations for input impedance and voltage gain. However, for low frequencies (below about 100 kHz) the error is minimal. If *only d-c feedback* is required, then the source resistor R_S is bypassed. With a capacitor across

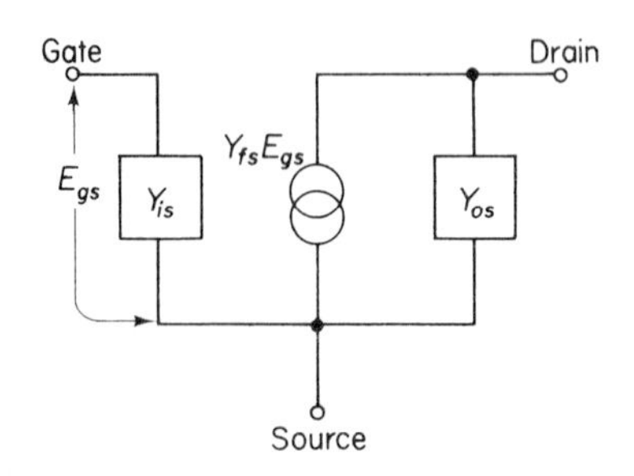

Fig. 2-27 Small signal analysis of basic FET stage (Courtesy of Motorola Inc., Semiconductor Products Division)

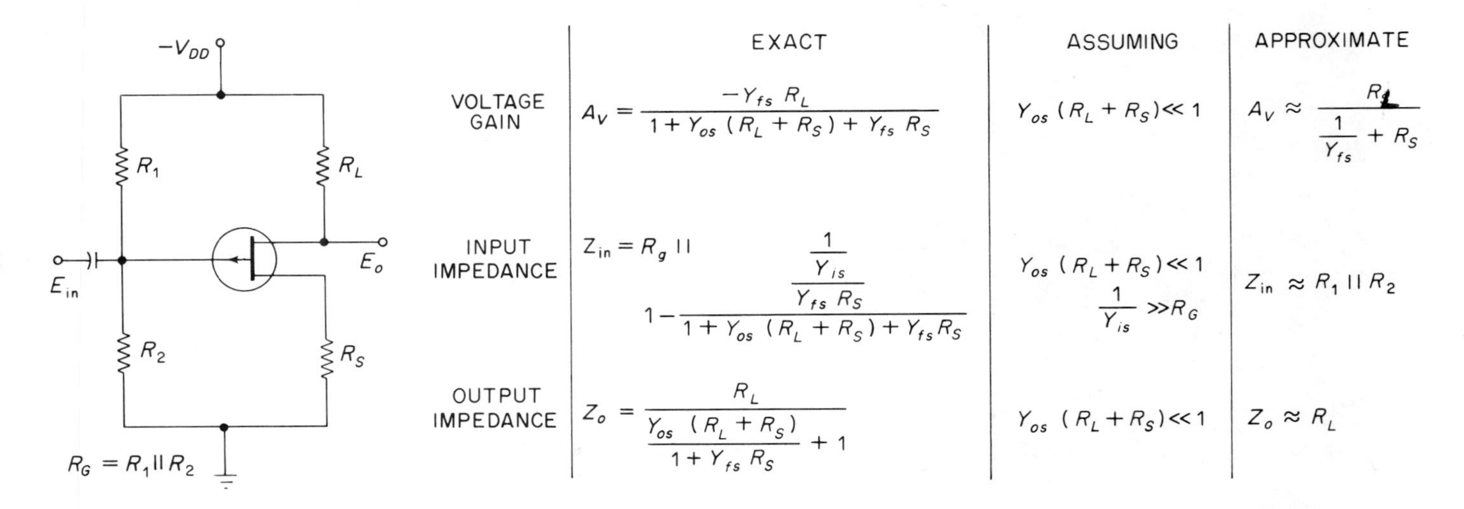

	EXACT	ASSUMING	APPROXIMATE
VOLTAGE GAIN	$A_V = \dfrac{-Y_{fs} R_L}{1 + Y_{os}(R_L + R_S) + Y_{fs} R_S}$	$Y_{os}(R_L + R_S) \ll 1$	$A_V \approx \dfrac{R_L}{\dfrac{1}{Y_{fs}} + R_S}$
INPUT IMPEDANCE	$Z_{in} = R_g \,\|\, \dfrac{\dfrac{1}{Y_{is}}}{1 - \dfrac{Y_{fs} R_S}{1 + Y_{os}(R_L + R_S) + Y_{fs} R_S}}$	$Y_{os}(R_L + R_S) \ll 1$ $\dfrac{1}{Y_{is}} \gg R_G$	$Z_{in} \approx R_1 \,\|\, R_2$
OUTPUT IMPEDANCE	$Z_o = \dfrac{R_L}{\dfrac{Y_{os}(R_L + R_S)}{1 + Y_{fs} R_S} + 1}$	$Y_{os}(R_L + R_S) \ll 1$	$Z_o \approx R_L$

Fig. 2-28 Basic common-source FET stage circuit and characteristics (Courtesy of Motorola Inc., Semiconductor Products Division)

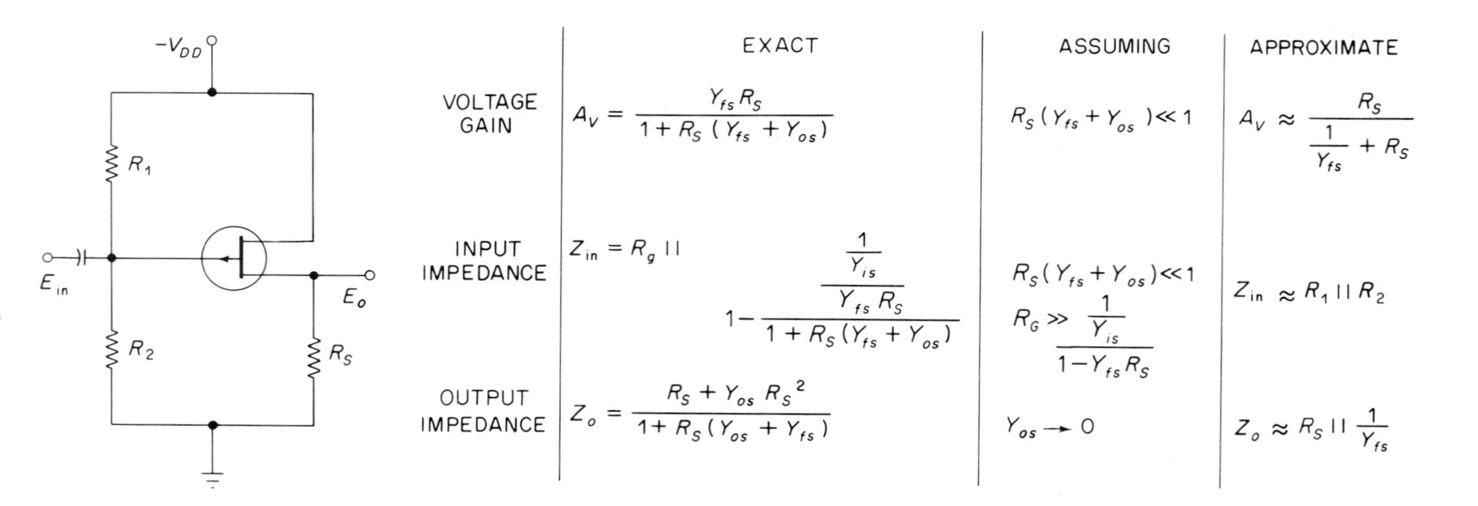

	EXACT	ASSUMING	APPROXIMATE
VOLTAGE GAIN	$A_V = \dfrac{Y_{fs} R_S}{1 + R_S\,(Y_{fs} + Y_{os})}$	$R_S\,(Y_{fs} + Y_{os}) \ll 1$	$A_V \approx \dfrac{R_S}{\dfrac{1}{Y_{fs}} + R_S}$
INPUT IMPEDANCE	$Z_{in} = R_g \,\|\, \dfrac{\dfrac{1}{Y_{is}}}{1 - \dfrac{Y_{fs} R_S}{1 + R_S (Y_{fs} + Y_{os})}}$	$R_S (Y_{fs} + Y_{os}) \ll 1$ $R_G \gg \dfrac{\dfrac{1}{Y_{is}}}{1 - Y_{fs} R_S}$	$Z_{in} \approx R_1 \,\|\, R_2$
OUTPUT IMPEDANCE	$Z_o = \dfrac{R_S + Y_{os} R_S^{\,2}}{1 + R_S (Y_{os} + Y_{fs})}$	$Y_{os} \rightarrow 0$	$Z_o \approx R_S \,\|\, \dfrac{1}{Y_{fs}}$

Fig. 2-29 Basic common-drain (source follower) FET stage circuit and characteristics (Courtesy of Motorola Inc., Semiconductor Products Division)

	EXACT	ASSUMING	APPROXIMATE
VOLTAGE GAIN	$A_V = \dfrac{Y_{fs}R_L}{1 + Y_{os}\,(R_L + R_S) + Y_{fs}R_S}$	$Y_{os}\,(R_L + R_S) << 1$	$A_V \approx \dfrac{R_L}{\dfrac{1}{Y_{fs}} + R_S}$
INPUT IMPEDANCE	$Z_{in} = \dfrac{1}{Y_{os} + Y_{fs}} + (R_S + R_L)\,\dfrac{Y_{os}}{Y_{os} + Y_{fs}} + \dfrac{R_S Y_{fs}}{Y_{os} + Y_{fs}}$	$Y_{os} << Y_{fs}$ $Y_{os}\,(R_L + R_S) << 1$	$Z_{in} \approx R_S + \dfrac{1}{Y_{fs}}$
OUTPUT IMPEDANCE	$Z_o = \dfrac{R_L}{\dfrac{Y_{os}\,(R_L + R_S)}{1 + Y_{fs}R_S} + 1}$	$Y_{os}\,(R_L + R_S) << 1$	$Z_o \approx R_L$

Fig. 2-30 Basic common-gate FET stage circuit and characteristics (Courtesy of Motorola Inc., Semiconductor Products Division)

R_S, the effects on the design equations are to set R_S at zero. Under these conditions, voltage gain is the product of R_L and Y_{fs}.

With R_S unbypassed, the circuit characteristics are virtually independent of FET parameters (with the exception of Y_{fs}). Instead, the circuit characteristics (impedances, gain, etc.) are dependent on R_L and R_S. By using precision resistors with close temperature coefficients the common-source circuit can be made very stable over a wide temperature range.

The common-drain (source-follower) configuration (Fig. 2-29) is a very useful basic circuit. Some of its properties are: a voltage gain always less than unity with no phase inversion, low output impedance (essentially set by the value of R_S), high input impedance, large signal swing, and active impedance transformation.

The common-gate stage (Fig. 2-30) offers impedance transformation opposite that of the source follower. Common-gate produces low input impedance and high output impedance. The voltage gain is the same as for the common-source stage in that there is no phase inversion.

The circuits of Figs. 2-28, 2-29, and 2-30 are for *P*-channel JFETs. Reverse the polarity for *N*-channel FETs. MOSFET (or IGFET) devices may not conform exactly to the relations shown, but are sufficiently close for first trial values.

2-10 Basic FET Amplifier Stage

Figure 2-31 is the working schematic of a basic, single-stage FET amplifier. Note that the basic amplifier circuit is similar to the basic common-source circuit of Figs. 2-25 and 2-28, except that input and output coupling capacitors C_1 and C_2 are added. These capacitors prevent direct-current flow to and from external circuits. Note that a bypass capacitor C_3 is shown connected across the source resistor R_S. Capacitor C_3 is required only under certain conditions, as discussed in Sec. 2-11.

Input to the amplifier is applied between gate and ground, across R_2. Output is taken across the drain and ground. The input signal adds to, or subtracts from, the bias voltage across R_2. Variations in bias voltage cause corresponding variations in I_D, and the voltage drop across R_L. Therefore, the drain voltage (or circuit output) follows the input signal waveform except that the output is inverted in phase. (If the input swings positive, the output swings negative, and vice versa.)

Variations in I_D also cause variations in voltage drop across R_S, and a change in the gate-source bias relationship. As discussed in Secs. 2-5 and 2-8, the change in bias that results from the voltage drop across R_S tends to cancel the initial bias change caused by the input signal, and serves as a form of *negative feedback* to increase stability (and limit gain). This form of gate-source feedback is known as *stage feedback* or *local feedback*, since only one

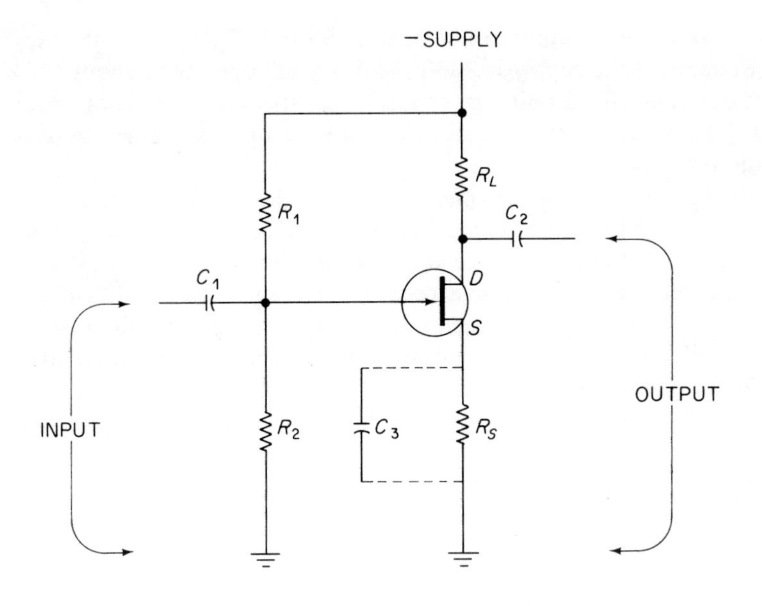

$$A_V \approx \frac{R_L}{\frac{1}{Y_{fs}} + R_S} \approx \frac{R_L}{R_S}$$

DRAIN VOLTAGE = 0.5 X SUPPLY

$$Z_{IN} \approx R_1 || R_2 \approx R_2$$

$$Z_{OUT} \approx R_L$$

Fig. 2-31 Basic common-source FET amplifier stage

stage is involved. As discussed in later sections, *overall feedback* or *loop feedback* is sometimes used where several stages are involved.

The outstanding characteristic of the circuit in Fig. 2-31 is that circuit characteristics (gain, stability, impedance) are determined primarily by circuit values, rather than by FET characteristics.

2-10-1 Design Considerations

The circuit as shown uses a *P*-channel FET. Reverse the power supply polarity if an *N*-channel FET is used.

If a maximum supply voltage is specified in the design problem, the maximum peak-to-peak output voltage is set. For class A operation, the drain is operated at approximately one-half the supply voltage. This permits the maximum positive and negative swing of output voltage. The peak-to-peak output voltage cannot exceed the source voltage. Generally, the absolute maximum peak-to-peak output can be between 90 and 95 per cent of the supply. For example, if the supply is 20 V, the drain will operate at about 10 V (quiescent point, no-signal point, or simply Q-point), and swing from about 1 V to 10 V. However, there is less distortion if the output is one-half to one-third of the supply voltage.

If a supply voltage is not specified, two major factors should determine the value: the maximum drain-source V_{DS} of the FET, and the desired output voltage [or the desired drain voltage at the operating point (Q-point)]. Obviously, the maximum V_{DS} cannot be exceeded. Preferably, the supply voltage should not exceed 90 per cent of the maximum V_{DS} rating. This allows a 10 per cent safety factor. Any desired output voltage (or drain Q-point voltage) can be selected within these limits.

If the circuit is to be battery-operated, choose a supply voltage that is a multiple of 1.5.

If a peak-to-peak output voltage is specified, add 10 per cent (to the peak-to-peak value) to find the *absolute minimum* supply voltage.

If a drain Q-point voltage is specified, double the drain Q-point voltage.

For minimum distortion, use a supply that is two to three times the desired output voltage.

If the input and/or output impedances are specified, the resistance values (R_1, R_2, R_L, R_S) are set, as shown in Fig. 2-31. However, there are certain limitations for R_2 and R_L imposed by tradeoffs (for gain, impedance match, zero-temperature-coefficient operating point, etc.).

For example, the output impedance is set by R_L. If R_L is increased to match a given impedance, the gain will increase (all other factors remaining equal). However, an increase in R_L will lower the drain voltage Q-point, since the same amount of I_D will flow through R_L and produce a larger voltage drop. This reduces the possible output voltage swing. A reduction in R_L has the opposite effect: increasing the drain voltage Q-point, but still reducing output voltage swing.

When R_1 is much larger than R_2 (which is generally the case), the input impedance of the circuit is set by the value of R_2. If R_2 is increased (or decreased) far from the value found in Sec. 2-8, the no-signal I_D point will change. For example, in Sec. 2-8 the I_D is set at 1 mA. If this value is the zero-temperature-coefficient (0TC) drain current and R_2 is changed drastically, the drain current will change and the FET is no longer operating at the 0TC point. Generally, this is the least desirable alternative. Typically, the common-source FET circuit is chosen for its high input impedance, thus presenting a low current drain to the preceding circuit. If the FET stage must provide a low input impedance, the common-gate circuit of Fig. 2-30 is generally preferred.

The values of coupling capacitors C_1 and C_2 are dependent upon the *low-frequency limit* at which the amplifier is to operate. As frequency increases,

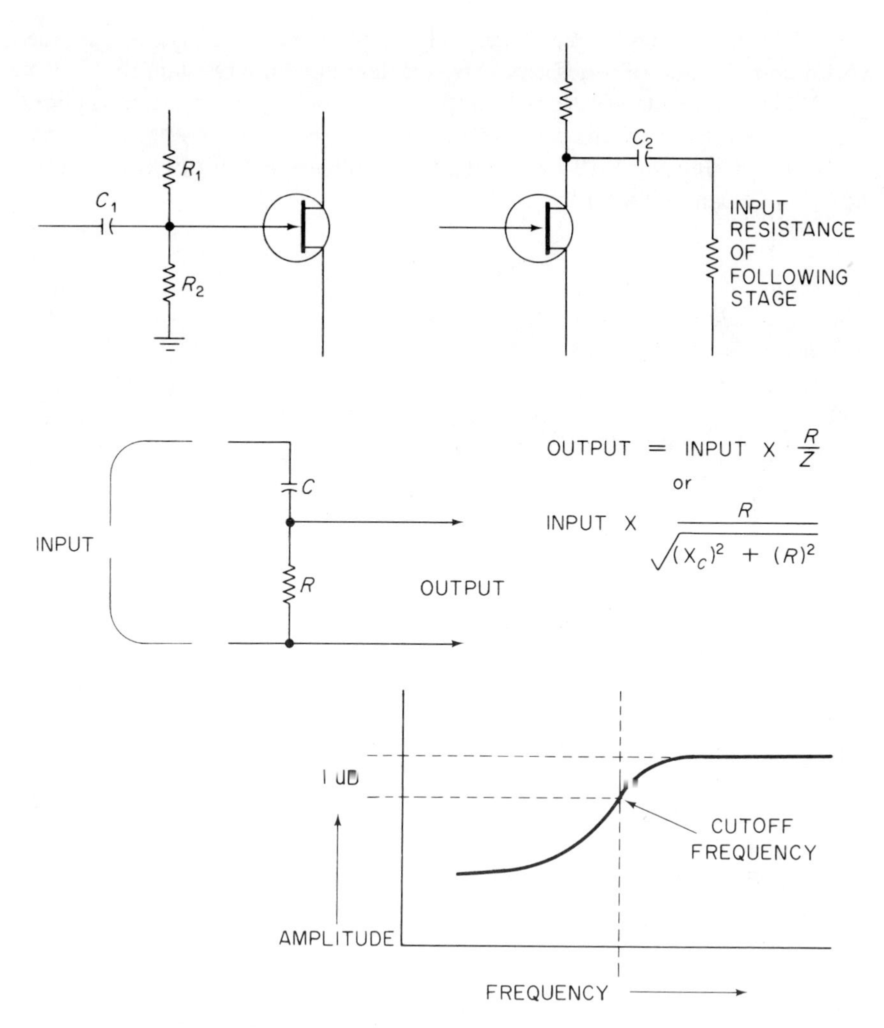

Fig. 2-32 Formation of high-pass *RC* filter by coupling capacitors and related resistances

capacitive reactance decreases and the coupling capacitor becomes (in effect) a short to the signal. Therefore, the high-frequency limit need not be considered in audio circuits. Capacitor C_1 forms a high-pass RC filter with R_2. Capacitor C_2 forms another high-pass filter with the input resistance of the following stage (or the load). This condition is shown in Fig. 2-32. The input voltage is applied across the capacitor and resistor in series. The output is taken across the resistance. The relation of input voltage to output voltage is:

$$\text{output voltage} = \text{input voltage} \times \frac{R}{Z}$$

where R is the d-c resistance value, and Z is the impedance obtained by the vector combination of series capacitive reactance and d-c resistance.

When the reactance drops to approximately one-half of the resistance, the output drops to approximately 90 per cent of the input (or approximately 1 dB loss). Using the 1-dB loss as the low-frequency cutoff point, the value of C_1 or C_2 can be found by:

$$\text{Capacitance} = \frac{1}{3.2\,FR}$$

where capacitance is in microfarads, F is the low-frequency limit in Hz, and R is resistance in megohms.

2-10-2 Design Example

Assume that the circuit of Fig. 2-31 is to be used as a single-stage voltage amplifier. The desired output is a minimum of 5 V (peak-to-peak) with an input of 1 V. This requires a gain of at least 5. The low-frequency limit is 30 Hz, with a high-frequency limit of 100 kHz. Minimum distortion is desired. (The circuit should not be overdriven.) The supply voltage is specified as -30 V. The FET type is specified as 2N5268. The I_D for zero-temperature-coefficient (0TC) is 1 mA. The minimum and maximum Y_{fs} are 2000 and 4000 micromhos, respectively. Input and output impedances are not specified, but must be calculated for reference to other circuits.

Calculate resistance values. The first step is to calculate the resistance values as described in Sec. 2-8. Since all factors are the same, the values are $R_1 = 9.1$ M, $R_2 = 300$ kΩ, $R_L = 16.9$ kΩ, $R_S = 2.2$ kΩ.

Supply voltage and operating point. With a supply voltage of 30 V and an output of 5 V, the ratio is much greater than the required 3-to-1 (or 2-to-1), so distortion should be at a minimum. With 1 mA of I_D flowing, the drop across R_1 is about 17 V, placing the drain at about 13 V. This drain voltage Q-point will easily permit a 5-V peak-to-peak output swing.

Minimum gain. Since a minimum gain of 5 is specified, use the minimum value of Y_{fs} (2000 μmhos) to find gain. Using the equation of Fig. 2-28, the gain is:

$$A_V \approx \frac{16.9\ \text{k}\Omega}{(1/2000\ \mu\text{mhos}) + 2.2\ \text{k}\Omega} = \frac{16.9\ \text{k}\Omega}{0.5\ \text{k}\Omega + 2.2\ \text{k}\Omega}$$

$$= \frac{16.9\ \text{k}\Omega}{2.7\ \text{k}\Omega} \simeq 6$$

A gain of 6 exceeds the required minimum of 5.

Note that the equation of Fig. 2-31 shows the approximate voltage gain is equal to R_L/R_S, without regard to Y_{fs}. This simplified rule holds true unless Y_{fs} and R_S are small. For example, if Y_{fs} is 800 μmhos and R_S 1250 ohms, the calculated gain will be double if Y_{fs} is ignored. Therefore, for a quick check of gain, use the R_L/R_S equation. If greater accuracy is desired, use the equation of Fig. 2-28.

Input impedance. Input impedance of the Fig. 2-31 circuit is the parallel combination of R_1 and R_2. However, since R_1 is many times (more than 10) the value of R_2, the approximate circuit input impedance is equal to R_2, or about 300 kΩ.

Output impedance. Output impedance of the Fig. 2-31 circuit is approximately equal to R_L, or 16.9 kΩ.

Coupling capacitors. The value of C_1 forms a high-pass filter with R_2. The high limit of 100 kHz can be ignored. The low-frequency limit of 30 Hz requires a capacitance value of:

$$C_1 \approx \frac{1}{3.2 \times 30 \times 0.3} \approx 0.03 \text{ to } 0.04\ \mu\text{F}$$

This will provide an approximate 1-dB drop at the low-frequency limit of 30 Hz. If a greater drop can be tolerated, the capacitance value of C_1 can be lowered. The value of C_2 is found in the same manner, except the resistance value must be the load resistance. The voltage values of C_1 and C_2 should be 1.5 times the maximum voltage involved, or $30 \times 1.5 = 45$ V.

Sufficient feedback. As a final check of the design values, compare the calculated gain versus the ratio of R_L/R_S. The gain should be at least 75 per cent of the resistance ratio. If so, there is sufficient feedback to be of practical value. In this example, the ratio is approximately 7.6, with the gain slightly over 6; 75 percent of 7.6 is approximately 5.7. Thus, the gain is greater, and there is sufficient feedback.

2-11 Basic FET Amplifier with Source Resistance Bypass

Figure 2-31 shows (in phantom) a bypass capacitor C_3 across source resistor R_S. This arrangement permits R_S to be removed from the circuit as far as the signal is concerned, but leaves R_S in the circuit (in regard to direct current). With R_S removed from the signal path, the voltage gain is approximately equal to $Y_{fs} \times R_L$. Thus, the use of a bypass capacitor permits a temperature-stable d-c circuit to remain intact, while providing a high signal gain.

2-11-1 Design Considerations

A source resistance bypass capacitor also creates some problems. The Y_{fs} changes with frequency, and from FET to FET. Thus, circuit gain can only be approximated. The source bypass is recommended where maximum voltage gain must be obtained from a single stage.

The value of C_3 can be found by:

$$\text{capacitance} = \frac{1}{6.2F(R_S \times 0.2)}$$

where capacitance is in microfarads, F is low-frequency limit in Hz, and R_S is in megohms.

2-11-2 Design Example

Assume that C_3 is to be used as a source bypass for the circuit described in the previous design example (Sec. 2-10-2) to increase voltage gain. All of the circuit values remain the same, as does the low-frequency limit of 30 Hz. Assume that the transistor has a Y_{fs} of 2000 μmhos minimum (same as previous example), and that the same 5-V minimum output is required.

When C_3 is added, the equation for approximate voltage gain changes to:

$$A_V \approx Y_{fs} \times R_L \qquad \text{or} \qquad 0.002 \times 16.9 \approx 33.8$$

With this gain, the input can be reduced to about 150 mV (from the 1 V in the previous example) to produce the same 5-V output ($5/33.8 \approx 0.15$). Keep in mind that the same transistor (2N5268) might have a Y_{fs} of 4000 μmhos. If so, the stage voltage gain will be about 68, and the 5-V output can be accomplished with about 75 mV. Likewise, if the 150-mV input is applied to the circuit with a 4000-μmho Y_{fs} transistor, the output will be over 10 V. This output is safe as far as distortion and clipping are concerned, since the drain voltage Q-point is about 13 V; however, the output may be excessive for the load or stage following the circuit. These factors should be considered before using source bypass.

The low-frequency limit of 30 Hz requires a C_3 capacitance value of:

$$C_3 \approx \frac{1}{6.2 \times 30 \times 0.0022 \times 0.2} \approx 12\mu\text{F}$$

This value provides a reactance across R_S that is about one-fifth of R_S, and effectively shorts the source (signal path) to ground. The voltage value of C_3 should be 1.5 times the maximum voltage involved, or 45 V.

2-12 Basic FET Amplifier with Partially Bypassed Source

Figure 2-33 is the working schematic of a basic, single-stage, FET amplifier with a partially bypassed source resistor. This design is a compromise between the basic design without bypass (Sec. 2-10) and the fully bypassed source (Sec. 2-11). The direct-current characteristics of both the unbypassed

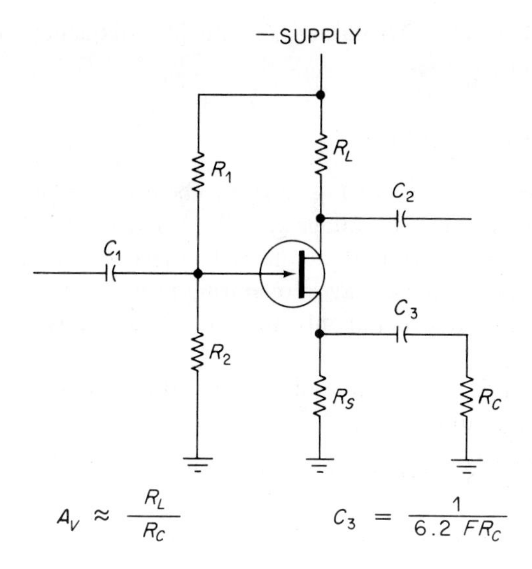

Fig. 2-33 Basic FET audio amplifier with partially bypassed source resistor

and partially bypassed circuits are essentially the same. All circuit values (except C_3 and R_C) can be calculated in the same way for both circuits.

As shown in Fig. 2-33, the voltage gain for a partially bypassed FET amplifier is greater than the unbypassed circuit, but less than for the fully bypassed circuit. However, the gain can be set to an approximate value by selection of circuit value (unlike the fully bypassed circuit where gain is entirely subject to variations in Y_{fs}).

2-12-1 Design Considerations

Design considerations for the circuit of Fig. 2-33 are the same as those for Fig. 2-31 (Sec. 2-10) except for the effect of C_3 and R_C.

The value of R_C is chosen on the basis of desired voltage gain. R_C should be substantially smaller than R_S, otherwise there will be no advantage to the partially bypassed design. As shown by the equations, voltage gain is approximately equal to R_L/R_C. This holds true unless both Y_{fs} and R_C are very low (where $1/Y_{fs}$ is about equal to R_C). In such a case, a more accurate gain approximation is:

$$\frac{R_L}{(1/Y_{fs}) + R_C}$$

The value of C_3 is found by:

$$\text{capacitance} = \frac{1}{6.2FR_C}$$

where capacitance is in microfarads, F is the low-frequency limit in Hz, and R_C is in megohms.

2-12-2 Design Example

Assume that the circuit of Fig. 2-33 is to be used in place of the Fig. 2-31 circuit, and that the desired voltage gain is 12 (twice that of the unbypassed circuit, but less than half that of the fully bypassed circuit). Selection of component values, supply voltages, operating point, and the like, is the same for both circuits; therefore, the only difference in design is selection of values for C_3 and R_C.

The value of R_C is the value of R_L divided by the desired gain, less the reciprocal of minimum Y_{fs}, or:

$$\frac{16.9}{12} \approx 1400;\ 1400 - 500 \approx 900 \text{ ohms}$$

The low-frequency limit of 30 Hz requires a C_3 capacitance value of:

$$\frac{1}{6.2 \times 30 \times 0.0009} \approx 6\ \mu\text{F}$$

The voltage value of C_3 should be 1.5 times the maximum voltage involved, or 45 V.

2-13 Basic FET Source Follower (Common Drain)

Figure 2-34 is the working schematic of a basic single-stage FET source-follower (common-drain) circuit. Note that this circuit is similar to that of Fig. 2-29, except that input and output coupling capacitors C_1 and C_2 are added. These capacitors prevent direct-current flow to and from external circuits.

Input to the source follower is applied between gate and ground across R_2. Output is taken across the source and gound. The input signal adds to, or subtracts from, the bias voltage across R_S. Variations in bias voltage cause corresponding variations in I_D and the voltage drop across R_S. Therefore, the source voltage (or circuit output) follows the input signal waveform, and remains in phase.

Variations in voltage drop across R_S change the gate-source bias relationship. This change in bias tends to cancel the initial bias change caused by the input signal and serves as a form of negative feedback to increase stability and limit gain.

The circuit of Fig. 2-34 is used primarily where high input impedance and output impedance (with no phase inversion) are required, but no gain is needed. The source follower is the FET equivalent of the transistor emitter follower and the vacuum tube cathode follower.

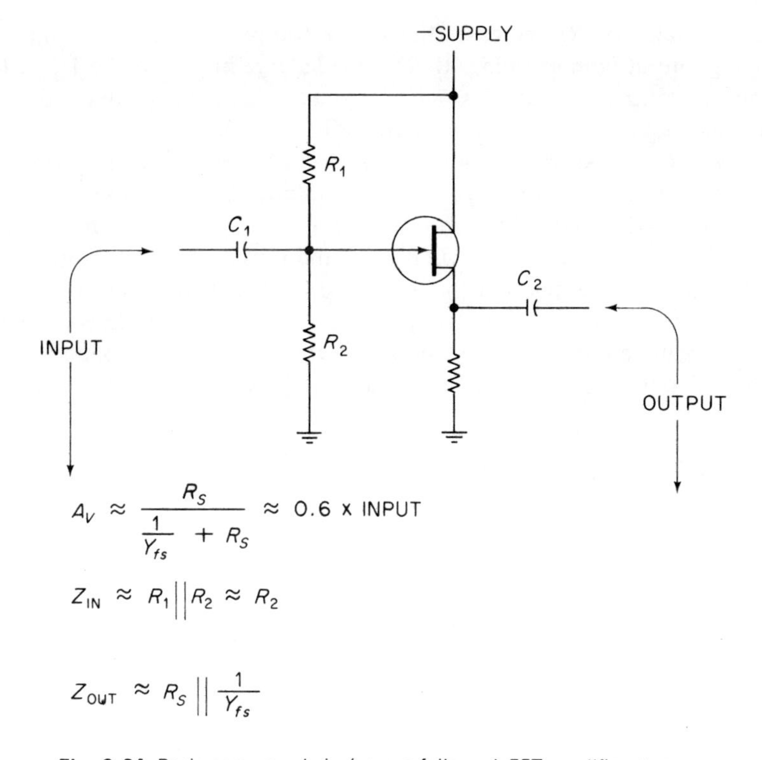

Fig. 2-34 Basic common-drain (source follower) FET amplifier stage

2-13-1 Design Considerations

The design considerations for the source follower of Fig. 2-34 are essentially the same as for the FET common-source amplifier of Sec. 2-10, with the following exceptions.

Output Q-point. The Q-point voltage at the circuit output (source terminal) is set by I_D under no-signal conditions, and the value of R_S. Since R_S is typically small, the Q-point voltage is quite low in comparison to the common-source amplifier. In turn, the maximum allowable peak-to-peak output voltage is also low. For example, if the source is at 1 V with no signal, the maximum possible peak-to-peak output is less than 1 V. Of course, the value of R_S can be increased as necessary to permit a higher output.

If the input and/or output impedances are specified in a design problem, the resistance values (R_1, R_2, R_S) are set, as shown in Fig. 2-34. However, there are certain limitations for R_S imposed by tradeoffs (for impedance match and output Q-point).

For example, the output impedance is the parallel resistance combination of R_S and $1/Y_{fs}$. If R_S is made very small (less than 10 times) in relation to $1/Y_{fs}$, then the output impedance is approximately equal to R_S. As discussed, a low value of R_S decreases the source (output) voltage Q-point, thus reducing output voltage swing. If R_S is made large in relation to $1/Y_{fs}$, then the output impedance is approximately equal to $1/Y_{fs}$, and is subject to variation with frequency, and from FET to FET.

Current gain. As shown by the equation of Fig. 2-34, there is no voltage gain for a source follower. Typically, the output voltage is about 0.6 times the input voltage, depending upon the ratio of $1/Y_{fs}$ to R_S. However, the source follower is capable of current gain, and thus *power gain*. For example, assume that 1 V is applied at the input and 0.6 V is taken from the output. Further assume that the input impedance is 300 kΩ, and the output impedance is 300 ohms. The input power is approximately 0.0033 mW, while the output power is about 1.2 mW, indicating a power gain of about 350.

2-13-2 Design Example

Assume that the circuit of Fig. 2-34 is to be used as a single-stage source follower. The desired output is a minimum of 0.6 V (peak-to-peak) with an input of 1 V. The low-frequency limit is 30 Hz. All other conditions are the same as for the design example in Sec. 2-10-2.

Calculate resistance values. The first step is to calculate the resistance values as described in Sec. 2-8 and Sec. 2-10, omitting the calculation for R_L. Since all factors are the same, the values are $R_1 = 9.1$ MΩ, $R_2 = 300$ kΩ, $R_S = 2.2$ kΩ.

Operating point. With an I_D of 1 mA, and an R_S of 2.2 kΩ, the drop across R_S (and the operating point) is about 2.2 V. This source voltage Q-point will easily permit a 0.6-V peak-to-peak output swing.

Minimum ouput signal voltage. Since a minimum output of 0.6 V is specified with 1-V input, the stage loss factor must be no greater than 0.6. Use the minimum value of Y_{fs} (2000 μmhos) to find the loss factor. Using the equation of Fig. 2-34, the loss factor is:

$$\frac{2200}{(1/0.002) + 2200} \approx \frac{2200}{2700} \approx 0.7$$

Thus, output signal voltage should be about 0.7 V peak-to-peak.

Input impedance. The input impedance of the Fig. 2-34 circuit is the parallel combination of R_1 and R_2; however, since R_1 is many times (more than 10) the value of R_2, the approximate circuit input impedance is equal to R_2, or about 300 kΩ.

Output impedance. Output impedance of the Fig. 2-34 circuit is the parallel combination of R_S and $1/Y_{fs}$, or

$$\frac{2200 \times 500}{2200 + 500} \approx 400 \text{ ohms}$$

Coupling capacitors. The values of the coupling capacitors C_1 and C_2 are the same as for the example of Sec. 2-10-2.

2-14 Basic FET Common Gate

Figure 2-35 is the working schematic of a basic, single-stage FET common-gate amplifier. Note that the basic circuit is similar to the basic common-gate circuit of Fig. 2-30, except that input and output coupling capacitors C_1 and C_2 are added. These capacitors prevent direct-current flow to and from external circuits.

Input to the common-gate amplifier is applied at the source across a portion of R_S. Typically, the value of R_{S1} is equal to R_{S2}, although it may be

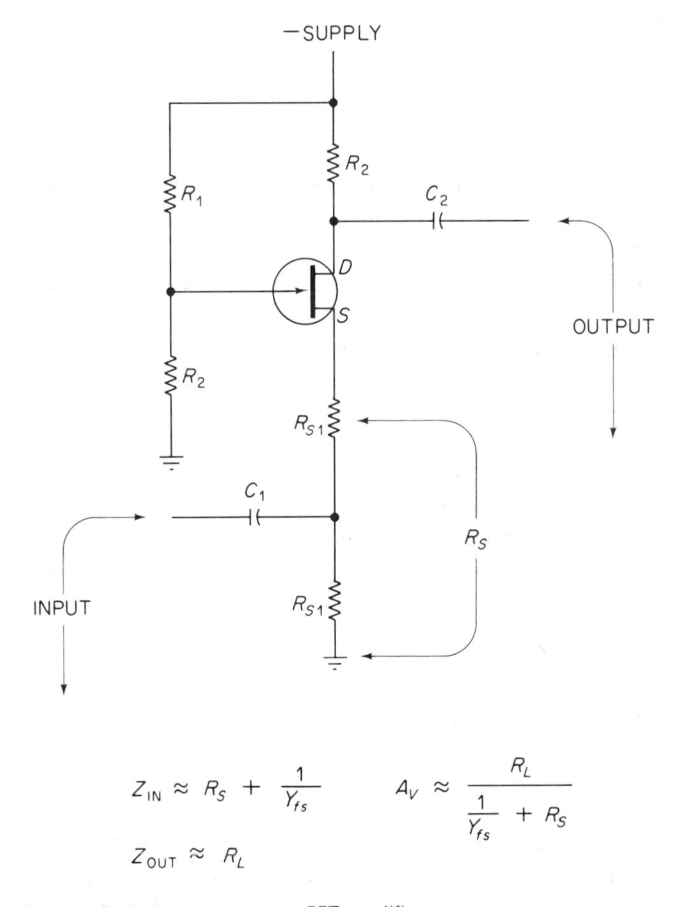

$$Z_{IN} \approx R_S + \frac{1}{Y_{fs}} \qquad A_V \approx \frac{R_L}{\frac{1}{Y_{fs}} + R_S}$$

$$Z_{OUT} \approx R_L$$

Fig. 2-35 Basic common-gate FET amplifier stage

necessary to divide the resistance value unequally. In any event, the total value of $R_S(R_{S1} + R_{S2})$ must be considered when calculating the direct-current characteristics of the circuit. Output is taken across the drain and ground. The input signal adds to, or subtracts from, the bias voltage across R_S. Variations in bias voltage cause corresponding variations in I_D and the voltage drop across R_L. The drain voltage (or circuit output) follows the input signal in phase.

Variations in voltage drop across R_S change the gate-source bias relationship. This change in bias tends to cancel the initial bias change caused by the input signal and serves as a form of negative feedback to increase stability (and limit gain).

The circuit of Fig. 2-35 is used primarily where low input impedance and high output impedance (with no phase inversion) are required. Gain is determined primarily by circuit values rather than by FET characteristics. The common-gate amplifier is the FET equivalent of the transistor common-base amplifier and the vacuum tube common-grid amplifier.

2-14-1 Design Considerations

Design considerations for the common-gate amplifier of Fig. 2-35 are essentially the same as for the FET common-source amplifier of Sec. 2-10, with the following exceptions.

If the input and/or output impedances are specified in a design problem, the resistance values (R_1, R_2, R_L, R_S) are set, as shown in Fig. 2-35. However, note that input impedance of the circuit is dependent upon the reciprocal of Y_{fs} $(1/Y_{fs})$ factor. This is true unless the value of R_S is many times (at least 10) that of $1/Y_{fs}$.

The values of coupling capacitors C_1 and C_2 are dependent upon the low-frequency limit at which the amplifier is to operate. Capacitor C_1 forms a high-pass RC filter with R_{S2}. Capacitor C_2 forms another high-pass filter with the input resistance of the following stage (or the load).

Using a 1-dB loss as the low-frequency cutoff point, the value of C_1 can be found by:

$$\text{capacitance} = \frac{1}{3.2FR_{S2}}$$

where capacitance is in microfarads, F is the low-frequency limit in Hz, and R_{S2} is in megohms.

2-14-2 Design Example

Assume that the circuit of Fig. 2-35 is to be used as a single-stage common-gate amplifier. All conditions are the same as for the design example in Sec. 2-10-2, with the following exceptions.

Calculate resistance values. The first step is to calculate resistance values as described in Secs. 2-8 and 2-10. Since all factors are the same, the values are $R_1 = 9.1\ \text{M}\Omega$, $R_2 = 300\ \text{k}\Omega$, $R_L = 16.9\ \text{k}\Omega$, $R_S = 2.2\ \text{k}\Omega$.

Note that the total value of R_{S1} and R_{S2} must equal 2.2 kΩ. As a first trial, use 1.1 K for both R_{S1} and R_{S2}.

Input impedance. Input impedance of the Fig. 2-35 circuit is the combination of $R_S + 1/Y_{fs}$. With a minimum Y_{fs} of 2000 μmhos and a maximum Y_{fs} of 4000 μmhos, the $1/Y_{fs}$ factors are 500 and 250, respectively. Thus, maximum input impedance is $2200 + 500 = 2.7\ \text{k}\Omega$, with minimum input impedance of $2200 + 250 = 2450$ ohms.

Coupling capacitors. The value of C_1 forms a high-pass filter with R_{S2}. The low-frequency limit of 30 Hz requires a capacitance value of

$$C_1 \approx \frac{1}{3.2 \times 30 \times 0.0011} \approx 10\ \mu\text{F}$$

This will provide an approximate 1-dB drop at the low-frequency limit of 30 Hz. If a greater drop can be tolerated, the capacitance value of C_1 can be lowered. The value of C_2 is found in the same manner, except the resistance value must be the load resistance.

2-15 Basic FET Amplifier without Fixed Bias

Figure 2-36 is the working schematic of a basic single-stage FET amplifier, without fixed bias. Note that the circuit is similar to the basic self-bias circuit of Fig. 2-9, except that the input and output coupling capacitors C_1 and C_2 are added, as is resistor R_1. The capacitors prevent direct-current flow to and from external circuits. The resistor R_1 provides a path for bias and signal voltages between gate and source.

Input to the amplifier is applied between gate and ground, across R_1. Output is taken across the drain and ground. The input signal adds to, or subtracts from, the bias voltage across R_1. Variations in bias voltage cause corresponding variations in I_D and the voltage drop across R_L. Therefore, the drain voltage (or circuit output) follows the input signal waveform, except that the output is inverted in phase.

Variations in I_D also cause variation in voltage drop across R_S, and a change in the gate-source bias relationship. The change in bias that results from the voltage drop across R_S tends to cancel the initial bias change caused by the input signal and serves as a form of negative feedback to increase stability (and limit gain).

The major difference in the circuit of Fig. 2-36 and the FET amplifier with fixed bias is that the amount of I_D at the Q-point is set entirely by the value of R_S. It may not be possible to achieve a desired I_D with a practical value of R_S. Thus, it may not be possible to operate at the zero-temperature-

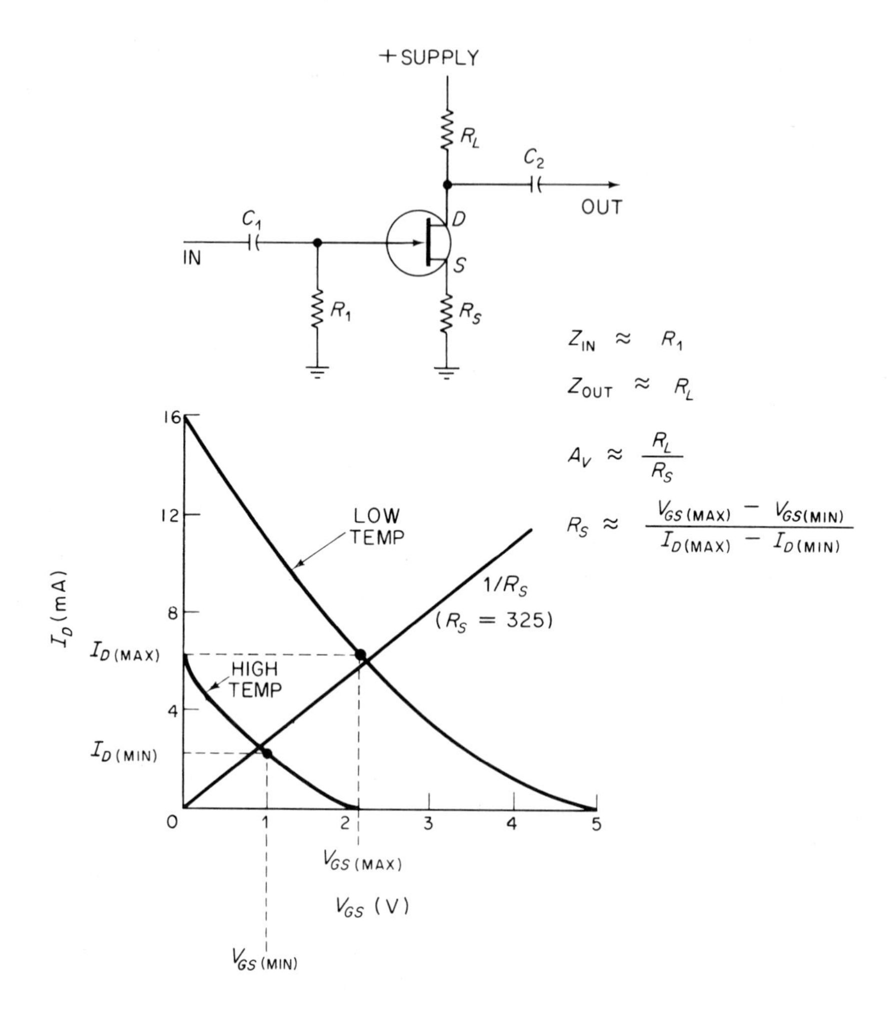

Fig. 2-36 Basic FET amplifier without fixed bias

coefficient point. If this is of less importance than minimizing the number of circuit components (elimination of one resistor), the circuit of Fig. 2-36 can be used in place of the fixed-bias FET amplifier.

2-15-1 Design Considerations

Design considerations for the amplifier of Fig. 2-36 are essentially the same as for the fixed-bias amplifier of Sec. 2-10-1, with the following exceptions.

If the input and/or output impedances are specified, the resistance values (R_1, R_S, R_L) are set, as shown in Fig. 2-36. However, there are certain limitations for the resistance values imposed by tradeoffs (for gain, impedance match, operating point, etc.).

For example, the value of R_S sets the amount of bias, and thus the amount of I_D. At the same time, the ratio of R_L/R_S sets the amount of gain. Going further, the value of R_L sets the output impedance. If R_S is changed in order to change the I_D, both the gain and drain Q-point will be changed. If R_L is changed to match a given impedance, both the gain and Q-point will change.

The input impedance is set by R_1. A change in R_1 will have little effect on gain, operating point, or output impedance. However, R_1 forms a high-pass RC filter with C_1. A decrease in R_1 requires a corresponding increase in C_1 to accommodate the same low-frequency cutoff point. As a general rule, the value of R_1 is high (in the megohm range). This minimizes current drain on the stage ahead of the FET.

2-15-2 Design Example

Assume that the circuit of Fig. 2-36 is to be used as a single-stage voltage amplifier without fixed bias. The desired output is 5 V (peak-to-peak) with an input of 1 V. This requires a gain of at least 5. The FET to be used has a minimum Y_{fs} of 2000 μmhos (a $1/Y_{fs}$ of 500), and V-I characteristics similar to those of Fig. 2-36. The desired I_D at the Q-point is 4 mA with an allowable swing from 2 to 6 mA. The supply is 30 V. The input impedance is specified as 3 megohms. The output impedance is unspecified, but must be calculated for reference to other circuits. The low-frequency limit is 30 Hz.

Calculate resistance values. The value of R_1 is set by the desired input impedance of 3 megohms. The value of R_S is determined by drawing a $1/R_S$ load line on the characteristics of Fig. 2-36, or by the equation, or both. Either way, the value of R_S is set by the limits of V_{GS} and I_D. The value of $V_{GS\,\text{min}}$ is the point where the $I_{D\,\text{min}}$ of 2 mA crosses the high-temperature limit curve, or approximately 1 V. The value of $V_{GS\,\text{max}}$ is the point where the $I_{D\,\text{max}}$ of 6 mA crosses the low-temperature curve, or approximately 2.3 V. Using these values, the first trial value for R_S is

$$R_S \approx \frac{2.3 - 1}{6 - 2} \approx 325 \text{ ohms}$$

Draw a $1/R_S$ line, starting from the 0 V_G and 0 I_D points, to make sure that it is possible for a straight line to pass between the two points, as shown in Fig. 2-36. With the $1/R_S$ line passing through the approximate middle of the two points, any point on the line represents approximately 325 ohms (when the corresponding voltage is divided by the corresponding current).

With the value of R_S set at 325 ohms and a $1/Y_{fs}$ of 500, the value of R_L must be at least 5 times (preferably more) greater to obtain the desired gain of 5. ($325 + 500 = 825 \times 5 = 4125$. Use 4.2 k$\Omega$.)

With R_L at 4.2 kΩ and 4 mA flowing at the Q-point, the drop across R_L is 16.8 V, which is approximately one-half the supply of 30 V. The drain voltage Q-point of approximately 13.2 V will permit a 5-V peak-to-peak output swing. Likewise, the 6-to-1 ratio of 5-V output to 30-V supply should be sufficient to keep distortion at a minimum. The output impedance of the circuit is approximately 4.2 kΩ (the value of R_L).

Coupling capacitors. The value of C_1 is:

$$C_1 \approx \frac{1}{3.2 \times 30 \times 3} \approx 0.003 \text{ to } 0.004\ \mu\text{F}$$

Sufficient feedback. As a final check of the design values, compare the calculated gain versus the ratio of R_L/R_S. For a highly stable circuit, the gain should be at least 75 per cent. In this example, the ratio is approximately 13, with the gain slightly over 5; 75 per cent of 13 is about 8. Thus the gain is less than 75 per cent, and there is insufficient feedback for a truly stable circuit. Of course, this circuit could be adequate for many applications.

There are two possible solutions to the problem of insufficient feedback. First, use a FET with a higher Y_{fs} value (lower $1/Y_{fs}$ factor). This makes R_S the dominant factor in the gain equation. The other solution is to use a much larger value of R_S. Of course, it may not be possible to use substantially larger R_S and still keep the I_D near the desired limits. From a practical standpoint, the best solution is to use some fixed bias.

2-16 Basic FET Amplifier with Zero Bias

Figure 2-37 is the working schematic of a basic single-stage FET amplifier, operating at zero bias and without feedback. Note that the circuit is similar to the basic circuit of Fig. 2-8, except that the input and output coupling capacitors C_1 and C_2 are added, as is resistor R_1. The capacitors prevent direct-current flow to and from external circuits. The resistor R_1 provides a path for signal voltages between gate and source.

Input to the amplifier is applied between gate and ground, across R_1. Output is taken across the drain and ground. The input signal varies the voltage on the gate. Variations in gate voltage cause corresponding variations in I_D and the voltage drop across R_L. Therefore, the drain voltage (or circuit output) follows the input signal waveform, except that the output is inverted in phase.

Any FET will have some value of I_D at zero V_{GS}. If the FET has characteristics similar to those of Fig. 2-37, the I_D will vary between about 1.75 mA and 4 mA, depending upon temperature, and from FET to FET. Therefore, with a zero-bias circuit, it is impossible to set the Q-point I_D at any particular value. Likewise, drain voltage Q-point is subject to considerable variation.

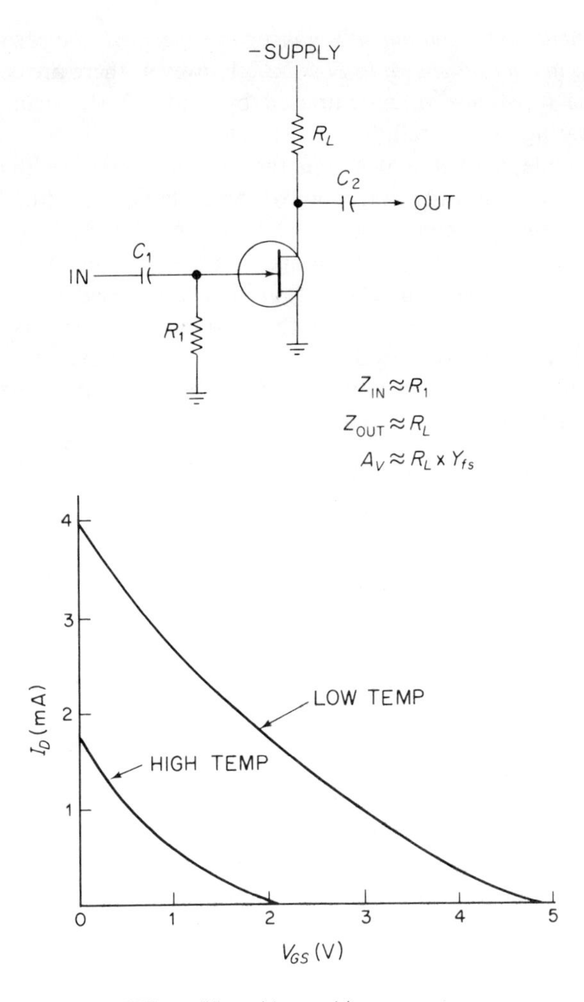

Fig. 2-37 Basic FET amplifier with zero bias

Since there is no source resistor, there is no negative feedback. Thus, there is no means to control this variation in I_D. For these reasons, the zero-bias circuit is used where circuit stability is of no particular concern.

The gain of a zero-bias circuit is set by Y_{fs} and the value of R_L, as shown by the equation of Fig. 2-37.

2-16-1 Design Considerations

Design considerations for the amplifier of Fig. 2-37 are somewhat similar to the fixed-bias amplifier of Sec. 2-10-1, but with the following exceptions.

If the input and/or output impedances are specified, the resistance values of R_1 and R_L are set, as shown in Fig. 2-37. However, there are certain limitations for the resistance values imposed by tradeoffs (for gain, impedance match, operating point, etc.).

For example, the value of R_L sets the amount of gain (with a stable Y_{fs}) and the drain voltage operating point (with a stable I_D). Thus, a change in Y_{fs} (which is usually accompanied by a change in I_D) causes a change in gain (and probably a shift in Q-point voltage). At best, the zero-bias circuit is unstable, even though the input and output impedances remain fairly constant. In designing the zero-bias circuit, both the minimum and maximum values of I_D must be considered, as well as the minimum and maximum Y_{fs}. If the final circuit meets the design requirements at both extremes, the trial values should be satisfactory.

2-16-2 Design Example

Assume that the circuit of Fig. 2-37 is to be used as a single-stage voltage amplifier without bias or feebdack. The desired output is 5 V (peak-to-peak) with an input of 0.5 V. This requires a gain of at least 10. The FET to be used has a minimum Y_{fs} of 3000 μmhos (0.003 mho) and a maximum Y_{fs} of 4000 μmhos (0.004 mho). The *V-I* characteristics are similar to those of Fig. 2-37. The I_D varies between 1.75 mA and 4 mA, depending on temperature, and from FET to FET. The supply is 30 V. The input impedance is specified as 3 megohms. The output impedance is unspecified but must be calculated for reference to other circuits. The low-frequency limit is 30 Hz.

Operating point. With a supply of 30 V, the operating point should be about one-half, or 15 V. Since it is impossible to determine the exact value of I_D at any given time, a compromise must be reached. Start by assuming that the voltage drop across R_L must be greater than the minimum peak-to-peak output voltage. With a minimum output of 5 V, assume that the minimum drop across R_L must be 7 V.

With a minimum I_D of 1.75 mA and a drop of 7 V, the value of R_L is 4 kΩ. With R_L at 4 kΩ and a maximum I_D of 4 mA, the drop across R_L is 16 V. Either 7 V or 16 V operating points are adequate for the 5-V output swing.

Gain. With R_L at 4 kΩ and a minimum Y_{fs} of 0.003, the gain is 12 (4000 $\times$ 0.003 = 12). With R_L at 4 kΩ and a maximum Y_{fs} of 0.004, the gain is 16. Either gain is sufficient to raise 0.5 V to a level greater than 5 V. Therefore, the basic gain requirements are met. However, assuming that the gain is 16 and the input is 0.5 V, the output will be 8 V (peak-to-peak). This may or may not cause distortion (or clipping) if the drain is at the 7-V *Q*-point. In practice, an 8-V peak-to-peak signal output should vary about 4 V on either side of the operating point, or from 3 to 11 V around a 7-V operating point.

The same output signal will vary from about 12 to 20 V, around the 16-V operating point.

Input impedance and coupling capacitors. The input impedance of 3 megohms is set by the value of R_1. The value of C_1 is:

$$C_1 \approx \frac{1}{3.2 \times 30 \times 3} \approx 0.003 \text{ to } 0.004\ \mu\text{F}$$

2-17 Multistage FET Amplifiers

In theory, any number of FET amplifier stages can be connected in cascade (output of one amplifier to input of next amplifier) to increase voltage gain. In practice, the number of stages is usually limited to three. The overall gain of the amplifier is the cumulative gain of each stage, multiplied by the gain of the adjacent stage. For example, if each stage of a three-stage amplifier has a gain of 10, the overall gain is 1000 ($10 \times 10 \times 10$). Since it is possible to design a fairly stable single FET stage with a gain of 15 to 20, a three-stage FET amplifier could provide gains in the 3000 to 8000 range. Generally, this is more than enough gain for most practical applications.

2-17-1 *Design Considerations for* JFET *Multistage Amplifiers*

Any of the single-stage FET amplifiers described in previous sections of this chapter could be connected together to form a two-stage or three-stage voltage amplifier. For example, the basic stage of Sec. 2-10 (with a gain of 5) can be connected to two like stages in cascade. The result is a highly temperature-stable voltage amplifier with a minimum gain of 125. Since each stage has its own feedback, the gain is precisely controlled and very stable.

It is also possible to mix stages to achieve some given design goal. For example, a three-stage amplifier can be designed by using the amplifier of Sec. 2-10 (gain of 5) for one stage, and the amplifier of Sec. 2-11 (gain of 30) for the remaining two stages. This results in an overall gain of 4500. Of course, with bypassed source resistors, the gain is dependent upon Y_{fs} and is therefore unpredictable. However, once the gain is established for a given amplifier, the gain should remain fairly stable.

Since design of a multistage JFET amplifier is essentially the same design for individual stages, no specific design example is given. In practical terms, design each stage as described in previous sections of this chapter, then connect the stages together. However, a few precautions must be considered.

Distortion and clipping. As is the case with any high-gain amplifier, be careful not to overdrive the circuit. If the maximum input signal is known, check this value against the overall gain and the maximum allowable output signal swing. For example, assume an overall gain of 1000 and a supply voltage of 20 V. With the final stage receiving a 20-V supply, the drain voltage *Q*-point is typically 10 V. Theoretically, this will allow a 20-V (peak-to-peak) output swing (from 0 to 20 V). In practice, a swing from about 1 to 19 V is more realistic. Either way, a 20-mV (*P-P*) input signal, multiplied by a gain of 1000, will drive the final output to its limits and possibly into distortion or clipping.

This problem brings up the tradeoff between controlled-gain stages (Sec. 2-10) and uncontrolled-gain stages (Sec. 2-11). If a minimum of gain is most important, use the uncontrolled-gain stages (or use two uncontrolled-gain stages to one controlled-gain stage). If a given output for a given input is the goal, use all controlled-gain stages.

Feedback. When each stage of a multistage amplifier has its own feedback (local or stage feedback), the most precise control of gain is obtained, However, such feedback is often unnecessary. Instead, overall feedback (or loop feedback) can be used, where part of the output from one stage is fed back to the input of a previous stage. Usually, such feedback is through resistance (to set the amount of feedback), and the feedback is from the final stage to the first stage. However, it is possible to use feedback from one stage to the next (second stage to first stage, third stage to second stage, etc.).

Be careful to watch for phase inversion when using loop or overall feedback. In a FET common-source amplifier, the phase is inverted from input (gate) to output (drain). If feedback is between two stages, the phase is inverted twice, resulting in *positive feedback*. This usually produces oscillation. In any event, it will not stabilize gain. If feedback between two stages must be used, connect the output (drain) of the second stage back to the source terminal of the first stage. This will produce the desired *negative feedback*. For example, if the gate (input) of the first stage is swinging positive, the drain of that stage will swing negative, as will the gate of the second stage. The drain of the second stage will swing positive, and this positive swing can be fed back to the source of the first stage. A positive input at the source of a common-source amplifier has the same effect as a negative at the gate. Thus, negative feedback is obtained.

Low frequency cutoff. When JFETs are connected to form multistage amplifiers, coupling capacitors must be used between stages, as well as at the input and output. If the coupling capacitors are omitted (*direct coupled*), the gate of each stage is at the d-c level of the previous drain. In a practical application, this results in some unrealistic value of I_D, so coupling capacitors are used to prevent d-c at the drain (or other input) from reaching the gate of the following stage.

As discussed in previous sections of this chapter, a coupling capacitor forms a low-pass RC filter with the gate-to-ground resistance. The effects of these filters are cumulative. For example, if each filter causes a 1-dB drop at some given cutoff frequency, and there are three filters (one at the input and two between stages), the result is a 3-dB drop at that frequency in the final output. If this cannot be tolerated, the RC relationship must be redesigned. In practical terms, this means increasing the value of C, since a change in R will usually produce some undesired shift in operating point or other circuit characteristic.

The following is a very approximate rule of thumb for finding the value of coupling capacitors with three stages involved:

$$C \approx \frac{1}{FR}$$

where C is in microfarads, F is the low-frequency cutoff (1 dB down) in Hz, and R is the gate-to-ground resistance in megohms.

This equation applies to the three coupling capacitors (one at the input of the first stage and two between stages), but does not take into account the capacitor at the drain of the output stage. The value of this capacitor must be calculated on the basis of whatever resistance follows the amplifier (the final load).

2-17-2 Design Considerations for IGFET *Multistage Amplifiers*

The considerations for multistage use of IGFETs are essentially the same for JFETs, with one major exception. Since the gate of an IGFET acts essentially as a capacitor, rather than a diode junction, no coupling capacitor is needed between stages. For a-c signals, this means that there is no low-frequency cutoff problem, in theory. In practical design, the input capacitance can form an RC filter with the source resistance and result in some low-frequency attenuation.

Figure 2-38 is the working schematic of an all-IGFET, three-stage amplifier. Note that all three IGFETs are of the same type, and all three drain resistors (R_1, R_2, R_3) are the same value. This simplifies design. At first glance, it may appear that all three stages are operating at zero bias. However, when I_D flows, there is some drop across the corresponding drain resistor, producing some voltage at the drain of the stage, and an identical voltage at the gate of the next stage. The gate of the first stage is at the same voltage as the drain of the last stage, because of feedback resistor R_F. There is no current drain through R_F, with the posible exception of reverse gate current (which can be ignored).

Operating point. To find a suitable operating point for the amplifier, it is necessary to trade off between desired output voltage swing, IGFET characteristics, and supply voltage. For example, assume that an output swing of

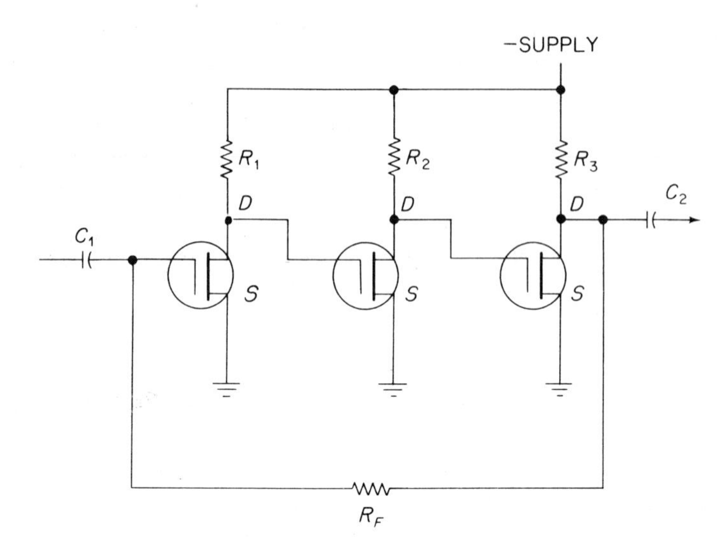

Fig. 2-38 Basic IGFET multistage amplifier

7 V peak-to-peak is desired, and the supply voltage is 24 V. Further assume that the I_D is about 0.5 mA when V_{GS} is 7 V. A suitable operating point would be 7 V to accommodate the 7-V output swing without distortion. (The swing would be from about 3.5 to 10.5 V, about the 7-V point.) This requires an 18-V drop from the 24-V supply. With 0.55 mA I_D, and an 18-V drop, the values of R_1, R_2, and R_3 should be about 33,000.

Gain. The overall voltage gain is dependent upon the relationship of the gain without feedback, and the feedback resistance R_F. Gain without feedback is determined by Y_{fs} and the values of R_1, R_2, and R_3. For example, assuming a Y_{fs} of 1000 μmhos (0.001 mho), the gain of each stage is 33 (33,000 × 0.001 = 33). With each stage at a gain of 33, the overall gain (without feedback) is about 36,000.

To find the value of R_F, divide the gain without feedback by the desired gain. Multiply the product by 100. Then multiply the resultant product by the value of R_1. For example, assume a desired gain of 3000. The gain without feedback is 36,000 (36,000/3000 = 12; 12 × 100 = 1200; 1200 × 33,000 = 39.6 MΩ, (use 40 MΩ).

Input impedance. The input impedance is dependent upon the relationship of gain and feedback resistance. The approximate input impedance is:

$$Z_{in} \approx \frac{R_F}{\text{gain}}$$

Since gain is dependent upon Y_{fs}, Z_{in} is subject to temperature, as well as to variation from FET to FET.

Direct-current amplifier. Note that the circuit of Fig. 2-38 requires one coupling capacitor at the input. This is necessary to isolate the input gate from any direct-current voltage that may appear at the input generator or other device. This makes the circuit of Fig. 2-38 unsuitable for use as a direct-current amplifier. The coupling capacitor forms an *RC* filter with the input resistance. However, since the resistance is so high, a 0.01-μF coupling capacitor will produce less than a 1-dB drop even at frequencies of a few Hz.

The circuit can be converted to a direct-current amplifier when the coupling capacitor is replaced by a series resistor R_{in}, as shown in Fig. 2-39.

The considerations concerning operating point are the same for both circuits. However, the series resistances must be terminated at a d-c level equivalent to the operating point. For example, if the operating point is -7 V, point *A* must be at -7 V. If point *A* is at some other d-c level, the operating point is shifted.

The relationships of Z_{in}, R_F, and gain still hold. However, Z_{in} is approximately equal to R_{in}. Therefore, gain is approximately equal to the ratio of R_F/R_{in}. This makes it possible to control the gain, by setting the ratio of R_F/R_{in}. Of course, the gain cannot exceed the gain-without-feedback (open loop) no matter what the ratio of R_F/R_{in} (closed loop). As a general rule, the greater the ratio of open-loop gain to closed-loop gain, the greater the circuit stability.

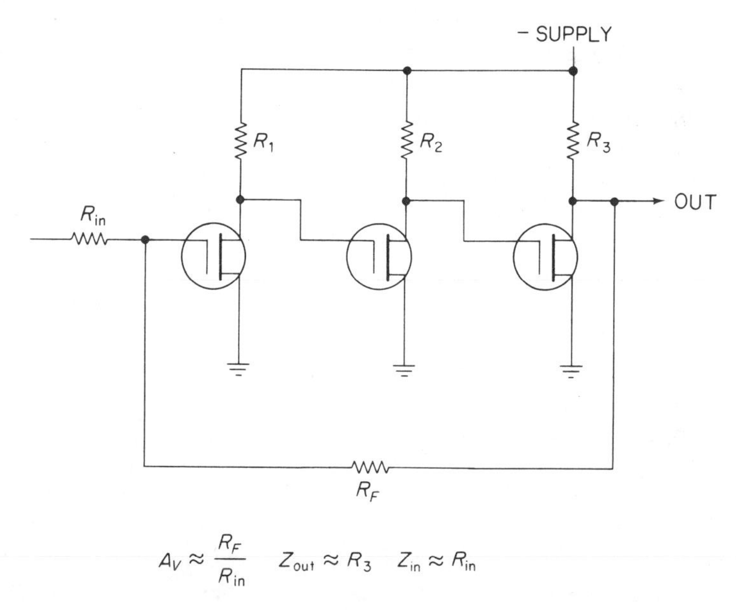

$$A_V \approx \frac{R_F}{R_{in}} \qquad Z_{out} \approx R_3 \qquad Z_{in} \approx R_{in}$$

Fig. 2-39 Basic IGFET direct-current amplifier

As an example, assume that the open-loop gain is 36,000 and the desired gain is 6000. This requires a ratio of 6 to 1. An R_F of 40 MΩ divided by 6 is 6.6 MΩ.

As a further example, assume that the desired gain is 5000, R_F is 40 MΩ, and the open-loop gain is 36 K. The desired gain of 5000 is considerably less than 36,000, so the circuit is very capable of producing the desired gain, with feedback. To find the value of R_{in}:

$$R_{in} \approx \frac{40{,}000{,}000}{5000} \approx 8000 \text{ ohms.}$$

2-18 Hybrid FET Amplifiers

In certain applications, FET stages can be combined with bipolar transistor stages to form hybrid amplifiers. The classic example is where a single FET stage is used at the input, followed by two bipolar amplifier stages. Such an arrangement takes advantage of both the FET and bipolar transistor characteristics.

A FET is essentially a voltage-operated device, permitting large voltage swings with low currents. This makes it possible to use high resistance values (resulting in high impedance) at the input and between stages. In turn, these high resistance values permit the use of low-value coupling capacitors, and eliminate the need for bulky, expensive electrolytic capacitors. If operated at the 0TC point, the FET is highly temperature stable, tending to make the overall amplifier equally stable. However, FETs have the characteristic of operating at low currents, and are therefore considered as low-power devices.

Bipolar transistors are essentially current-operated devices, permitting large currents at about the same voltage levels as the FET. Thus, with equal supply voltages and signal voltage swings, the bipolar transistor can supply much more current gain (and power gain) than the FET. Since currents are high, the impedances (input, interstage, output) must be low in bipolar transistor amplifiers. This requires large-value coupling capacitors, if low frequencies are involved. The low impedances also place a considerable load on devices feeding the amplifier, particularly if the devices are high impedance. On the other hand, a low output impedance is often a desirable characteristic for an amplifier.

When a FET is used as the input stage, the amplifier input impedance is high. This places a small load on the signal source, and allows the use of a low-value input coupling capacitor (if required). If the FET is operated at the 0TC point, the amplifier input is temperature-stable. (Generally, the input stage is the most critical in regard to temperature stability.) When bipolar transistors are used as the output stages, the output impedance is low, and current gain (as well as power gain) is high.

Hybrid amplifiers can be direct-coupled or capacitor-coupled, depending upon requirements. The direct-coupled configuration offers the best low-frequency response, permits direct-current amplification, and is generally simpler (uses fewer components). The capacitor-coupled hybrid amplifier permits a more stable design and eliminates the voltage regulation problem common to all direct-coupled amplifiers. (That is, a direct-coupled amplifier cannot distinguish between changes in signal level and changes in power supply level.)

The FET can be combined with any of the classic two-stage bipolar transistor amplifier combinations. The two most common combinations are the Darlington pair (for no voltage gain, but high current gain and low output impedance) and the *NPN-PNP* common-emitter amplifier pair (for both voltage gain and current gain).

2-18-1 Direct-Coupled Hybrid Amplifier

Figure 2-40 is the working schematic of a direct-coupled amplifier using a FET input stage and a bipolar transistor pair as the output. Note that local

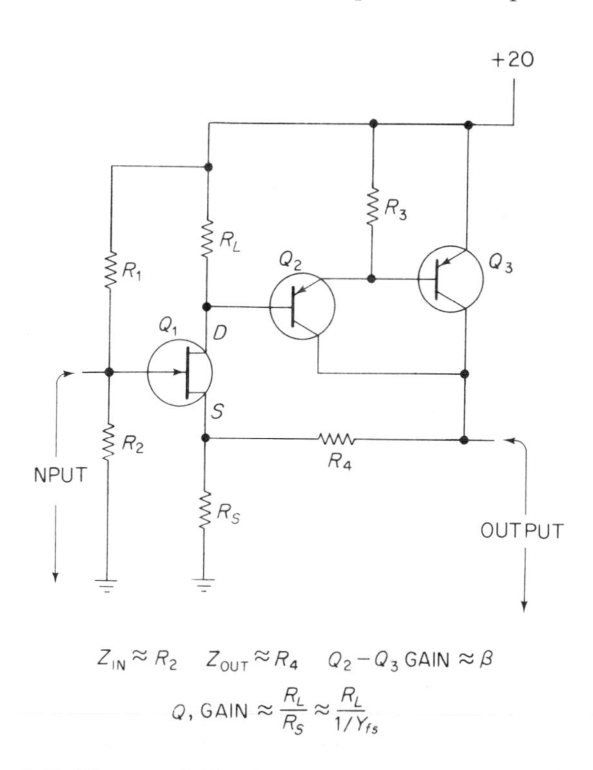

Fig. 2-40 Direct-coupled hybrid amplifier

feedback is used in the FET stage (provided by source resistor R_S), as well as overall feedback (provided by resistance R_4).

The design considerations for the FET portion of the circuit are essentially the same as described in previous sections, with certain exceptions. Input impedance is set by the value of R_2, as usual. Output impedance is set by the combination of R_4 and R_S. However, since R_S is quite small in comparison to R_4, the output impedance is essentially equal to R_4.

The gain of the FET stage is set by the ratio of R_L to R_S, plus the $1/Y_{fs}$ factor. However, since R_S is quite small, the FET gain is set primarily by the ratio of R_L to $1/Y_{fs}$. The gain of the bipolar transistor pair is set by the beta of the two transistors, and the feedback. Therefore, the gain can only be estimated.

Note that the drop across R_3 is the normal base-emitter drop of a transistor (about 0.5 to 0.7 V for silicon and 0.2 to 0.3 V for germanium). The drop across R_L is twice this value (about 1 to 1.5 V for silicon, 0.4 to 0.6 V for germanium). Thus, for a typical silicon transistor, the base of Q_2 and the drain of Q_1 operate at about 1 V removed from the supply. In a practical experimental circuit, R_L must be adjusted to give the correct bias for Q_2 (and operating point for Q_1). The same is true for R_3. However, as a first trial value, R_3 should be about twice the value of R_4.

Design starts with a selection of I_D for the FET. If maximum temperature stability is desired, use the 0TC level of I_D. This will probably require fixed bias, as described in Sec. 2-10. If temperature stability is not critical, the FET can be operated at zero bias by omitting R_1. There will be some voltage developed across R_S. However, since R_S is small, the V_{GS} is essentially zero, and the I_D is set by the zero V_{GS} characteristics of the FET.

With the value of I_D set, select a value of R_L that produces approximately 1- to 1.5-V drop, to bias Q_2.

The input impedance is set by R_2, with the output impedance set by R_4. The value of R_3 is approximately twice that of R_4. The value of R_S is less than 10 ohms, typically in the order of 3 to 5 ohms.

As a brief design example, assume that the circuit of Fig. 2-40 is to provide an input impedance of 1 megohm, an output impedance of 500 ohms, and maximum gain. Temperature stability is not critical.

Under these conditions, the values of R_2 and R_4 are set at 1 MΩ and 500 ohms (or the nearest standards). The value of R_S is 5 ohms, but the voltage drop across R_S can be ignored. FET Q_1 operates at $V_{GS} = 0$, for practical purposes. Assume that I_D is 0.2 mA under these conditions. With a required drop of 1.5 V and 0.2 mA I_D, the value of R_L is approximately 7.5 kΩ. Since R_4' is 500 ohms, R_3 should be 1 kΩ.

The key component in setting up this circuit is R_L. With the circuit operating in experimental form, adjust R_L for the desired Q-point voltage at the output (collectors of Q_2 and Q_3).

2-18-2 Capacitor-Coupled Hybrid Amplifier

Figure 2-41 is the working schematic of a capacitor-coupled hybrid amplifier. Only one capacitor is used, between the FET input stage and the transistor pair. Note that there is no overall feedback, but all three stages have local feedback.

The design considerations for the FET portion of the circuit are essentially the same as described in Sec. 2-10, except for the supply voltage of 20 V (30 V is used in the Sec. 2-10 example). This changes the values of R_1, R_2, R_L, and R_S, but the procedures for finding the values are the same and will not be repeated here.

Input impedance is set by the value of R_2. Output impedance is set by R_{L2}. The value of C_1 is set by the value of R_2 and the low-frequency limit. The value of C_2 is dependent upon the low-frequency limit and the value of R_B. The value of C_3 is dependent upon the low-frequency limit and the value of input resistance of the following stage (or the load).

The following is a summarized design example for the circuit of Fig. 2-41. Note that design for the FET portion of the circuit is the same as described in Secs. 2-8 and 2-10. Design for the transistor pair portion of the circuit is standard (using the author's techniques, as summarized here).

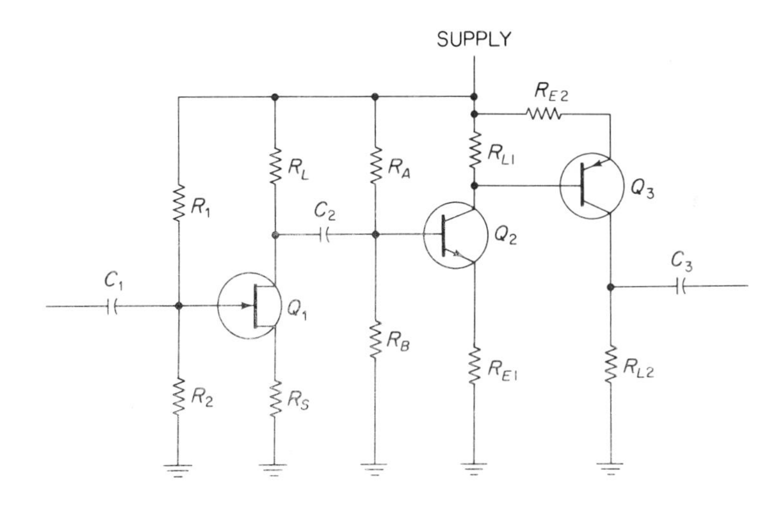

$Z_{IN} \approx R_2$

$Z_{OUT} \approx R_{L2}$

GAIN = SEE TEXT

Fig. 2-41 Capacitor-coupled hybrid amplifier

Assume that the circuit of Fig. 2-41 is to be used as a voltage amplifier. The desired output is 7 V (peak-to-peak) with a 500-ohm impedance. The input is 35 mV from a source in the order of 250 to 300 kΩ; 35-mV input and 7-V output requires an overall gain of 200. The low-frequency limit is 30 Hz. A source of 20 V is specified. Minimum distortion is desired. (The circuit should not be overdriven.) A 2N5268 FET is to be used for the input Q_1. Transistors Q_2 and Q_3 are 2N3568 and 2N3638, respectively.

FET portion of circuit. With a voltage gain of about 200 required, each stage must provide a gain of 6, if gain is divided equally between stages. It is better to let the FET provide high impedance and temperature stability, and keep gain at minimum. The transistor pair can provide most of the gain. If the FET provides a gain of 3, instead of 6, the transistor pair must provide a gain of 70, producing an overall gain of about 210. With an input of 35 mV and a gain of 3, the FET output should be about 100 mV.

Calculate resistance values. The first step is to calculate the FET resistance values as described in Sec. 2-8. With all factors the same except for the supply voltage of 20 V, the values are $R_1 = 6.3$ MΩ, $R_2 = 280$ kΩ, $R_L = 9.5$ kΩ, $R_S = 2.2$ kΩ.

Supply voltage and operating point. With a supply voltage of 20 V and an output of 100 mV, the ratio is much greater than the desired 3-to-1, so distortion should be at a minimum. With 1 mA of I_D flowing, the drop across R_1 is about 9.5 V, placing the drain at about 10.5 V. This drain voltage Q-point will easily permit a 100-mV peak-to-peak output swing.

Minimum gain. Since a minimum gain of 3 is specified for the FET stage, use the minimum value of Y_{fs} (2000 μmhos) to find gain. Using the equation of Fig. 2-28, the gain is:

$$A_V \approx \frac{9.5\text{k}\Omega}{(1/2000) + 2.2\text{ k}\Omega} \approx \frac{9.5\text{ k}\Omega}{2.7\text{ k}\Omega} \approx 3.5$$

Input impedance. Input impedance of the Fig. 2-41 circuit is the parallel combination of R_1 and R_2. However, since R_1 is many times (more than 10) the value of R_2, the approximate circuit input impedance is equal to R_2, or about 280 kΩ.

Coupling capacitor C_1. The value of C_1 is set by the value of R_2 and the low-frequency limit of 30 Hz, and is:

$$C_1 \approx \frac{1}{3.2 \times 30 \times 0.28} \approx 0.03\ \mu\text{F}$$

This will provide a drop of about 1 dB at the 30-Hz low limit. There will be corresponding drops at C_2 and C_3, resulting in an overall drop of about 3 dB at the 30-Hz point. Generally, an overall 3-dB drop is considered standard.

Sufficient gain. The ratio of R_L/R_S is about 4.1; 75 per cent of 4.1 is 3.08. The approximate gain is 3.5. Thus, gain is greater, and there is sufficient feedback.

Transistor pair portion of circuit. The transistor pair must provide a gain of approximately 70.

Supply voltage and operating point. The supply voltage, 20 V, is specified. Since this is approximately three times the desired output of 7 V, there should be no distortion.

First establish the Q-point for the output stage. The collector voltage should be approximately one-half the source, or $20/2 = 10$ V, at the Q-point.

Load resistance and collector current. The value of R_{L2} is specified as 500 ohms (to produce the desired output impedance). Use the nearest standard value of 510 ohms. With a 10-V drop across R_{L2}, the Q_3 collector current is 10/510 or 19.6 mA (rounded off to 20 mA).

Emitter resistance, current, and voltage. When two stages are direct-coupled in the transistor pair of Fig. 2-41, the overall voltage gain is about 70 per cent (or possibly more) of the combined gains of each stage. Since the required voltage gain is 70, the combined gains should be 100 (or a gain of 10 for each stage). To provide a voltage gain of 10 in the output stage, the value of R_{E2} should be one-tenth of R_{L2}, or $510/10 = 51$ ohms. The current through R_{E2} is the collector current of 20 mA, plus the base current. Assuming a circuit gain of 10, the base current is 20/10, or 2 mA. The combined currents through R_{E2} are $20 + 2$, 22 mA. This produces a drop of 1.12 V across R_{E2} (rounded off to 1 V).

Output base voltage and input collector voltage. The base of Q_3 should be 0.5 V from the emitter voltage. Since Q_3 is a *PNP*, the base should be more negative (less positive) than the emitter. The emitter of Q_3 is at +19 V (20 V − 1 V). Therefore, the base of Q_3 is +18.5 V. This sets the collector voltage for Q_2 at the Q-point.

Q_2 stage resistances and current. For the best signal transfer between Q_1 and Q_2, R_B should be the same value as R_L (or 9.5 kΩ). Use a value of 10 kΩ for R_B to simplify calculation. For maximum stability of the Q_2 stage, the ratio of R_B/R_{E1} should be 10. With R_B set at 10 kΩ, the value of R_{E1} is 1 kΩ. With the R_{E1} value established, and a stage voltage gain of 10 desired, the value of R_{L1} should be 1 kΩ × 10, or 10 kΩ.

With a 1.5-V drop across R_{L1}, the collector current is 1.5/10 kΩ = 0.15 mA. The current through R_{E1} is the collector current of 0.15 mA, plus the base current. Assuming a gain of 10, the base current is 0.15/10 = 0.015 mA. The combined currents through R_{E1} are 0.15 + 0.015 mA. This produces a drop of 0.165 mV across R_{E1}.

The base voltage of Q_2 is 0.5 V higher than the emitter voltage, or 0.5 V + 0.000165 = 0.500165 V. With a 0.500165-V drop across R_B, the current through R_B is 0.05 mA.

The value of R_A must be sufficient to drop the 20-V supply to about 0.5 V. The current through R_A is the current through R_B, 0.05 mA, plus the base current of 0.015 mA, or 0.065 mA. The resistance required to produce an approximate 19.5-V drop with 0.065 mA is 19.5/0.000065 = 300 kΩ. In a practical circuit, it is usually necessary to adjust R_A for the desired collector Q-point voltage at Q_3, rather than Q_2, with minimum distortion.

Coupling capacitors. The value of C_2 forms a high-pass filter with R_B. The low-frequency limit of 30 Hz requires a capacitor value of

$$C_2 \approx \frac{1}{3.2 \times 30 \times 10{,}000} \approx 1\ \mu\text{F}$$

The value of C_3 is found in the same manner, except the resistance value R must be the load resistance.

Transistor selection. In this case, the transistors were specified in the design example. However, it is always wise to check basic transistor characteristics against the proposed circuit parameters *before* making any connections.

The maximum collector voltages for 2N3568 and 2N3638 are 80 and 25 V, respectively, both well above the 20-V supply. Note that the FET stage could have been operated at 30 V, as described in Sec. 2-10. However, this is above the maximum for the 2N3638.

The power dissipated by Q_3 is about 200 mW (Q-point of 10 V times 20 mA). This is well below the 300-mW data-sheet limit. The dissipation of Q_2 is less than 2 mW.

Minimum betas for the 2N3568 and 2N3638 are 40 and 20, respectively, both well above the required 10 (for a stage gain of 10).

2-18-3 Nonblocking Amplifier

When an amplifier is required to have a coupling capacitor at the input, or between stages, it is possible that the amplifier will be *blocked* if JFETs or bipolar transistors are used. Either of these devices acts as a diode at the input, and serves to rectify the input signal. On one half-cycle, the diode is forward-biased, causing the capacitor to charge rapidly. On the other half-cycle, the diode is reverse-biased, causing the capacitor to discharge slowly. This can block the amplifier at certain frequencies, typically above 100 kHz. This problem can be eliminated by using IGFET at the input. The gate of an IGFET is similar to a capacitor. If a coupling capacitor is connected to an IGFET gate, the capacitor will charge and discharge at the same rate.

IGFET WARNING: It is assumed that the readers are familiar with IGFET handling procedures. IGFETs must be handled with care. In circuit, an IGFET is just as rugged as the JFET. Out of circuit, the IGFET is subject to damage from *static charges*. The IGFET is generally shipped with the leads all shorted together. Thus, there will be no static discharge between leads. Keep the leads shorted together whenever practical. When touching IGFET leads, keep your body at ground (keep one hand on chassis).

2-19 FET Schmitt Trigger

A Schmitt trigger circuit is a bistable network which switches from one stable state to the other when an input signal varies above or below predetermined voltage levels. The Schmitt trigger is widely used as a voltage-level detector in analog and analog-digital systems. The Schmitt trigger is also used as a pulse-shaping circuit.

A JFET can be used in conjunction with a bipolar transistor to form a Schmitt trigger. Such an arrangement takes advantage of the JFET's high input impedance and the bipolar transistor's high current capacity.

2-19-1 Design Considerations

Figure 2-42 shows a Schmitt trigger circuit configuration, using both the FET and bipolar transistors. Note that the input is applied at the FET gate, and the output is taken from the bipolar transistor collector. Assume, for

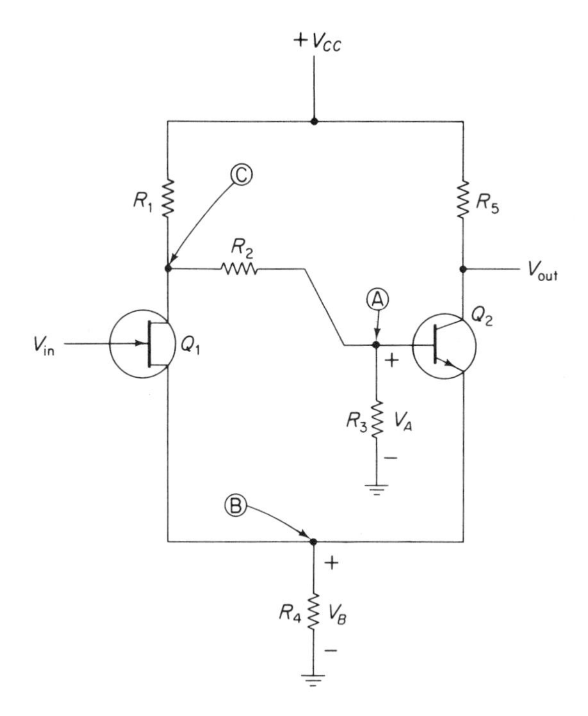

Fig. 2-42 Schmitt trigger circuits using FET at input (Courtesy, Texas Instruments Incorporated)

the present analysis, that the FET Q_1 is removed from the circuit. The voltage level at point A, V_A, is sufficiently positive to turn on transistor Q_2; and the voltage level at point B, V_B, is clamped by the emitter-base junction of Q_2 to one diode drop below V_A (typically 0.5 to 0.7 V for a silicon transistor). With FET Q_1 back in the circuit, the gate-source voltage V_{GS} of Q_1 is given by

$$V_{GS} = V_{\text{in}} - V_B \tag{2-1}$$

The FET is nonconducting if V_{GS} is at least as negative as the pinch-off voltage V_P of the device. Hence, from Eq. (2-1), the FET is pinched off if:

$$V_{\text{in}} - V_B \leq V_P \tag{2-2}$$

Circuit design is such that the inequality of Eq. (2-2) is valid for a zero-level input signal; output voltage of the circuit is low, as Q_2 is turned on and Q_1 is turned off. As V_{in} is made increasingly positive, a voltage level is reached where $V_{\text{in}} - V_B$ becomes slightly more positive than V_P and the FET begins to conduct current. The resulting voltage drop across R_1 causes Q_2 to partially turn off. This lowers the V_B level and Q_1 turns on even harder. The regenerative action causes the Schmitt trigger circuit to switch to the state where Q_1 is fully turned on and Q_2 is at cutoff. Output voltage V_{out} is then at a relatively positive level.

As V_{in} is reduced from a large positive voltage level toward ground potential, Q_1 begins to conduct less current. Eventually, a level of V_{in} is reached at which the inequality of Eq. (2-2) is satisfied. The circuit then switches to the state where Q_1 is turned off and Q_2 is fully on.

The quiescent states of the Schmitt trigger circuit correspond to the operating conditions where Q_1 is on and Q_2 is off, or where Q_1 is off and Q_2

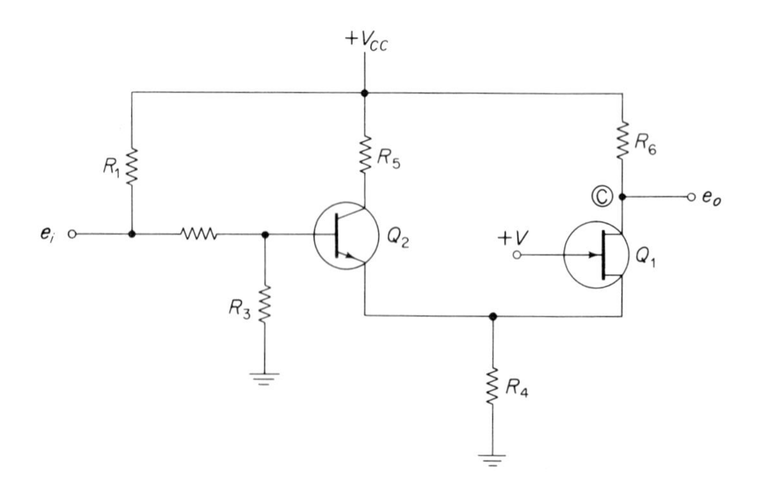

Fig. 2-43 Network for obtaining open-loop gain of Schmitt trigger circuit (Courtesy, Texas Instruments Incorporated)

is on. Therefore, the circuit has two stable states, determined by the level of V_{in}.

Open-loop gain. Regeneration occurs in the Schmitt trigger circuit when the open-loop gain of the circuit is equal to, or greater than, unity. This gain can be determined by opening the circuit at a convenient point and treating the resulting network as a linear amplifier. The circuit can be opened at the drain of Q_1 as shown in Fig. 2-43.

An input signal e_1 is applied to resistor R_2. The output signal e_0 appears at point C. Transistor Q_2 is biased in the active region by R_1 connected to the supply voltage. A d-c voltage +V is applied to the gate of Q_1 in order to bias Q_1 in the active region. An expression for open-loop gain can be described by

$$A_V \approx \frac{R_1 R_3}{\{[R_2 R_3 / R_E(h_{FE} + 1)] + R_2 + R_3\}[(1 + R_1 Y_{os})/(Y_{fs} + Y_{os})]} \tag{2-3}$$

where

A_V is open-loop voltage gain
Y_{os} is small-signal common-source output admittance of Q_1
Y_{fs} is small-signal common-source forward transfer admittance of Q_1
h_{FE} is d-c current gain of Q_2
R_E is equal to: $(1 + Y_{os}R_1)R_4/[R_4(Y_{fs} + Y_{os}) + R_1 Y_{os} + 1]$

Parameters Y_{os} and Y_{fs} in Eq. (2-3) are dependent upon d-c operating levels of the FET. Figure 2-44 shows measured plots of these two parameters for a 2N3824 FET. At extremely low current levels, the small value of Y_{fs} prevents regeneration in the Schmitt trigger circuit. However, as current increases in the FET, Y_{fs} becomes large enough for the open-loop gain to exceed unity.

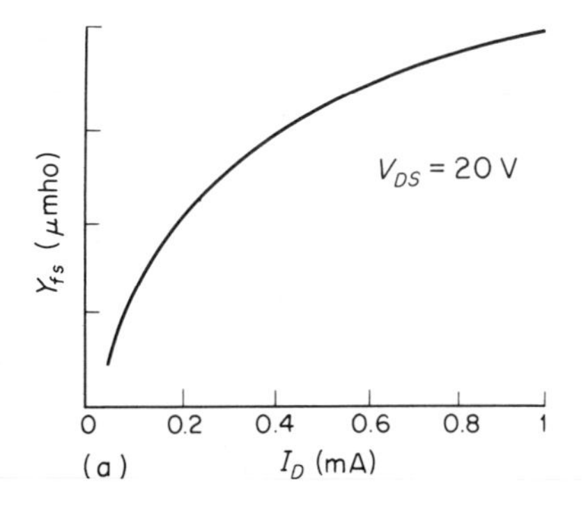

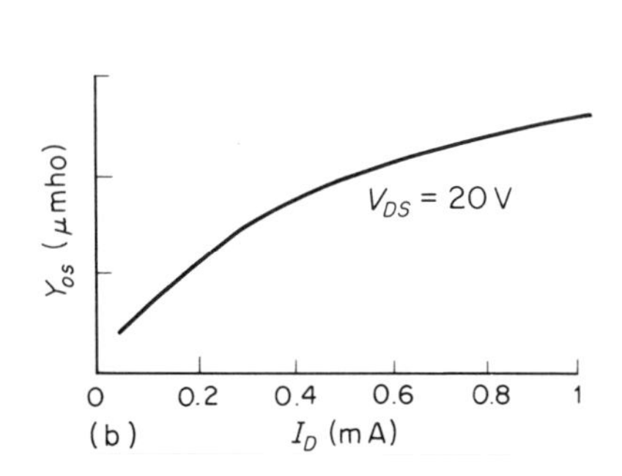

Fig. 2-44 Measured plots of (a) y_{fs} and (b) y_{os} for 2N3824 FET (Courtesy, Texas Instruments Incorporated)

Equation (2-3) can be reduced to fewer terms by using the relationships $R_1 Y_{os} \ll 1$ and $Y_{fs} \gg Y_{os}$. This allows Eq. (2-3) to be simplified to

$$A_V \approx \frac{R_1 R_3 Y_{fs}}{[R_2 R_3 / R_E(h_{FE} + 1)] + R_2 + R_3} \tag{2-4}$$

The above expression is used to determine Y_{fs} for unity circuit gain. From a Y_{fs}-Y_{GS} plot of the type shown in Fig. 2-45, the corresponding V_{GS} level can be determined.

The validity of Eq. (2-4) was checked by breadboarding the circuit of Fig. 2-46. The 2N3824 FET (the characteristics of which are shown in Figs. 2-44 and 2-45) is used in the circuit of Fig. 2-46. For a given level of I_D, a corresponding value for Y_{fs} is obtained from Fig. 2-46a. This Y_{fs} level is then substituted into Eq. (2-4), together with resistor and h_{FE} values. Figure 2-46b shows measured and calculated plots of A_V versus I_D. These plots show that Eq. (2-4) gives a close approximation to open-loop voltage gain.

Trigger-voltage levels. The input voltage level at which the Schmitt trigger circuit changes from the low-output voltage state to the high-output state is referred to as V_{ON}. The input voltage level at which the circuit switches back to the low-output voltage state is designated V_{OFF}.

Consider that input voltage V_{in} (to the circuit of Fig. 2-42) is at a low potential, and that Q_2 is conducting. Transistor Q_2 may be either in saturation or in the active region, depending upon the circuit design. If Q_2 is saturated, the circuit does not behave as a linear amplifier, and a small voltage change at point C has no effect upon collector current of Q_2. In order for collector current of Q_2 to change, drain current of Q_1 must increase to the level where voltage drop across R_1 causes Q_2 to turn off slightly. However, if Q_2 is not saturated, the circuit is a linear amplifier at any level of drain

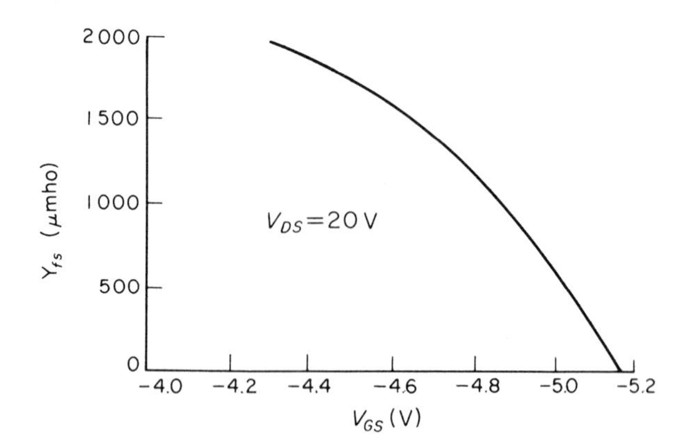

Fig. 2-45 Measured plot of y_{fs} versus V_{GS} for 2N3824 FET (Courtesy, Texas Instruments Incorporated)

current, and regeneration occurs at a lower level of V_{in}. Thus, Q_2 is maintained out of saturation for the following analysis:

For Q_1 at cutoff in Fig. 2-42, V_B is given by

$$V_B = V_A - V_{BE(ON)} \tag{2-5}$$

where: $V_{BE(ON)}$ is base-emitter forward voltage drop of Q_2. In the last expression, V_A can be described by

$$V_A = \frac{[V_{CC} - I_B(R_1 + R_2)]R_3}{R_1 + R_2 + R_3} \tag{2-6}$$

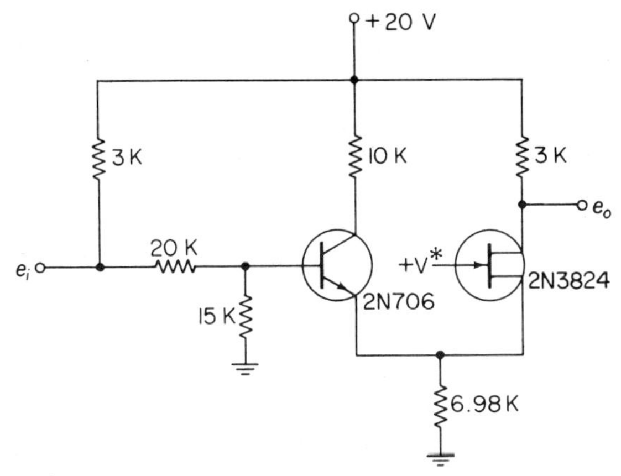

* +V is adjusted to give desired level of I_D

(a)

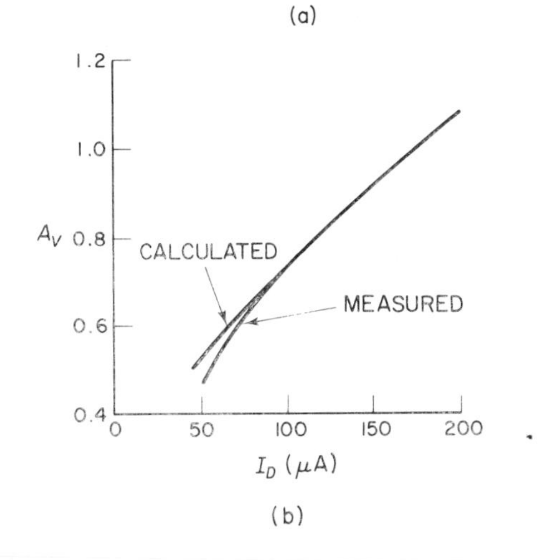

(b)

Fig. 2-46 (a) Test circuit for determining open-loop gain; (b) plots of measured and calculated open-loop gain (Courtesy, Texas Instruments Incorporated)

Substitute the right side of Eq. (2-5) for V_B in Eq. (2-1). Also, in Eq. (2-1), substitute V_{ON} for V_{in}. Solving the resulting expression for V_{ON} gives

$$V_{ON} = V_{GS} + V_A - V_{BE(ON)} \tag{2-7}$$

The latter equation gives the value of V_{ON} in terms of circuit parameters. Voltage V_{GS} is at the level which makes open-loop circuit gain equal unity.

When the Schmitt trigger circuit of Fig. 2-42 switches to the state where Q_1 is ON and Q_2 is OFF, base-emitter voltage of Q_2 is given by

$$V_{BE(OFF)} \frac{(V_{CC} - I_D R_1)R_3}{R_1 + R_2 + R_3} - I_D R_4 \tag{2-8}$$

where $V_{BE(OFF)}$ is off-state voltage of Q_2, and I_D is drain current of Q_1.

The above level of $V_{BE(OFF)}$ is now more negative than the base-emitter turn-on voltage of Q_2. As V_{in} is decreased, I_D decreases and $V_{BE(OFF)}$ becomes more positive. In order for Q_2 to begin conducting, the level of $V_{BE(OFF)}$ must increase to approximately 0.5 V (assuming that Q_2 is a low-power silicon transistor).

2-19-2 Design Example

Assume that the circuit of Fig. 2-42 is to be used as a Schmitt trigger. The circuit conditions are: $V_{ON} = 2$ V, $V_{OFF} = 1.8$ V, $V_{CC} = +20$ V. The FET is a 2N3824, having a V_P of approximately 5.15 V, and characteristics similar to those shown in Figs. 2-44 and 2-45. A 2N706 transistor is used for Q_2.

Collector current I_C and R_5. The product $I_C R_5$ determines voltage swing at the output. When Q_2 is conducting, the magnitude of V_{CC} is dropped across the series combination of R_5, Q_2, and R_4. At this point in the analysis, none of the above three voltage drops is known.

The voltage drop across R_4 can be described by combining Eqs. (2-5) and (2-7), and solving the resulting expression for V_B; this yields

$$V_B = V_{ON} - V_{GS} \tag{2-9}$$

When Q_1 turns ON, V_{GS} is slightly less negative than V_P (assume a V_{GS} of -5 V). Under these conditions, V_B is not larger than $2 - (-5)$, or 7 V.

At least 1 V should be maintained across $Q_2(V_{CE})$ in order to assure that Q_2 does not saturate. Voltage swing across R_5 can be as large as $20 - 7 - 1$, or 12 V. Select a value of 10 V for the product $I_C R_5$. This gives a V_{CE} level of 2 V for the ON state of Q_2. The choice of values for I_C and R_5 is dependent largely upon capacitive loading at the circuit output and any limitation upon fall time of the output voltage waveform (a lower value of R_5 will lower fall time). However, since this example has no specification for fall time, choose R_5 to be 10 kΩ. This gives an I_C level of 1 mA.

Resistors R_4 and R_3. For a collector current of 1 mA, V_{BE} of the 2N706 is measured to be 0.67 V. The value of V_{GS} to be used in the following equations is that value which will make open-loop gain equal to unity. However, since R_1, R_2, and R_3 are not known at this point, it is not possible to use Eq. (2-4) to determine the level of Y_{fs} (and, consequently, V_{GS}) to give the desired gain. As a first approximation, select V_{GS} from Fig. 2-45 to give a Y_{fs} of 500 μmhos (about -5 V). Substitution of the V_{GS} value (-5 V), the V_{BE} level (0.67 V), and the V_{ON} level (2 V) into the following equation yields a V_A of 7.67:

$$V_A = V_{ON} - V_{GS} + V_{BE} \tag{2-10}$$
$$= 2 - (-5) + 0.67$$

With V_A at 7.67, the voltage at point B(or V_B) is 7 V, since V_B is $V_A - V_{BE}$. The value of R_4 can be determined by

$$R_4 = \frac{V_B}{I_C} \tag{2-11}$$

Thus, $R_4 = 7/0.001 = 7\ \text{k}\Omega$.

Figure 2-47 shows the input circuit to the base of Q_2. For Q_2 turned off, and neglecting the small collector-base reverse leakage current I_B, voltage V_A is given by

$$V_A = I_0 R_3 \tag{2-12}$$

In order to maintain V_A at a level which is relatively independent of small changes in base current, let I_0 be 10 times the value of I_B. Common-emitter d-c current gain h_{FE} of the particular 2N706 transistor used here has a value of 82 at 1 mA of collector current and a V_{BE} of 0.67; 1 mA/82 gives an I_B level of 0.0122 mA. By making I_0 10 times I_B, the value of I_0 is 0.122 mA. With V_A at 7.67 and I_0 at 0.122, Eq. (2-12) shows that R_3 is 7.67/0.122 = 62.8 kΩ.

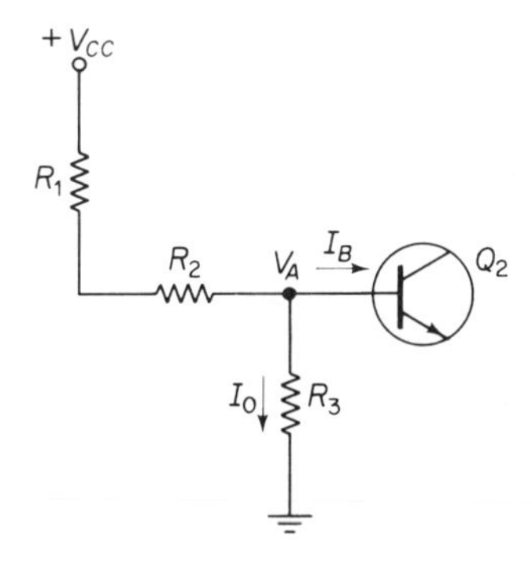

Fig. 2-47 Input circuit to base of Q_2 (Courtesy, Texas Instruments Incorporated)

Resistors R_1 and R_2. Current I_0 in the circuit of Fig. 2-47 can be described by

$$I_0 = \frac{V_{CC}}{1.1(R_1 + R_2) + R_3} \tag{2-13}$$

All terms except R_1 and R_2 are known at this point. Equation (2-8) also contains the terms R_1 and R_2. Equations (2-8) and (2-13) can be solved simultaneously for R_1 to give

$$R_1 = \frac{V_{CC}}{I_D} - \frac{(V_{BE(\text{OFF})} + I_D R_4)(V_{CC} + 0.1\, I_D R_3)}{1.1\, I_D I_0 R_3} \tag{2-14}$$

All terms of Eq. (2-14) are known, except $V_{BE(\text{OFF})}$ and I_D. $V_{BE(\text{OFF})}$ is assumed to be 0.5 V (which is logical for a silicon transistor). The value of I_D should be that current which is produced (by V_{in}) at the V_{OFF} level of 1.8 V.

A graphical procedure is used to determined I_D for a given level of V_{in}. A portion of the FET transfer curve is shown in Fig. 2-48, together with a load line having a slope of $1/R_4$. This load line intersects the horizontal axis at the V_{OFF} level of 1.8 V. The load line also intersects the I_D curve at about 0.875 mA. Thus, an I_D of 0.875 is flowing when the gate of Q_1 is at 1.8 V (the V_{OFF} level). Using Eq. (2-14), R_1 is found to be 4.2 kΩ.

With V_{CC}, I_0, R_1, and R_3 known, Eq. (2-13) can be rearranged to find R_2 as follows:

$$R_2 = \frac{(V_{CC}/I_0) - R_3}{1.1} - R_1 \tag{2-15}$$

$$\frac{(20/0.122\text{ mA}) - 62.8\text{ k}\Omega}{1.1} - 4.2\text{ k}\Omega = 87.8\text{ k}\Omega$$

2-19-3 Schmitt Trigger as Time-Delay Element

The high input impedance of the FET allows the hybrid Schimitt trigger circuit of Fig. 2-42 to be used as a long time-delay element. Figure 2-49 shows a circuit configuration for providing a time delay. Capacitor C is initially

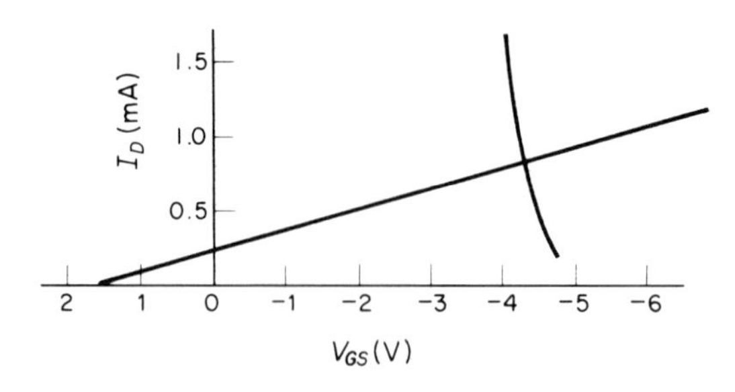

Fig. 2-48 Transfer curve and load line for design example (Courtesy, Texas Instruments Incorporated)

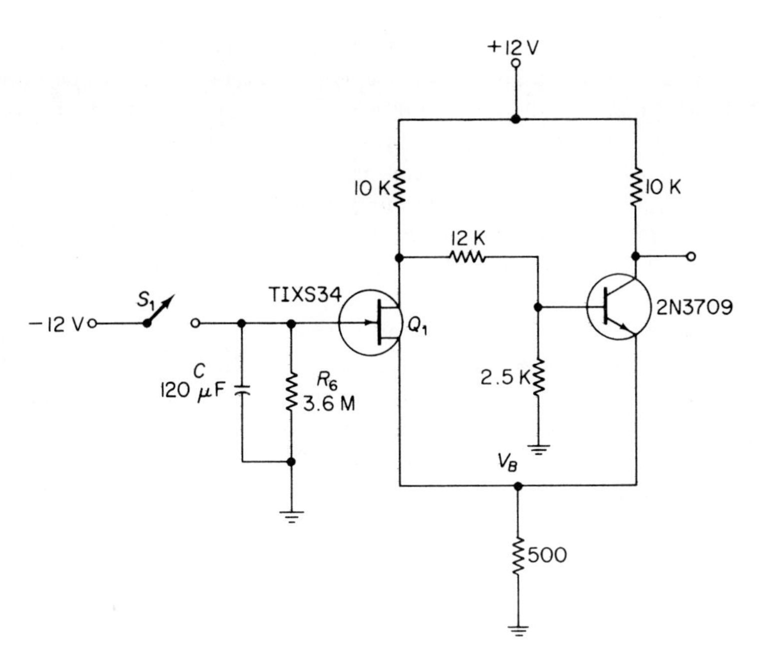

Fig. 2-49 Circuit for obtaining time delay of approximately 13 minutes (Courtesy, Texas Instruments Incorporated)

charged to −12 V by closing switch S_1. Gate-source voltage of Q_1 is given by $V_{GS} - 12 - V_B$.

The circuit is designed so that V_B is less negative than V_P of the FET. Under these conditions, the FET is pinched off because of the −12 V level at the gate terminal.

Switch S_1 is next opened and the capacitor C begins discharging through R_6 towards zero volts. Gate-source voltage of Q_1 can now be described by

$$V_{GS} = (-12e^{-t/R_6C}) - V_B$$

Eventually, V_{GS} of the FET becomes less negative than V_P, and the Schmitt trigger circuit changes state. A time delay of approximately 13 minutes is obtained from the circuit of Fig. 2-49 when Q_1 has a V_P level of +1.5 V. Using a FET having a smaller value for V_P will give even longer time delays.

2-20 FET Current Regulators

Field effect transistors are well adapted to circuits which require a constant current (or current regulation). Output impedances of FETs can range anywhere from the kilohm region to the tens of megohms, depending upon the device and configuration used. FETs are available with values of g_{os} (output conductance) ranging from one-half to several hundred micromhos. Saturation voltages, or the minimum operating voltages, are in the region of 1 V. The maximum voltage before breakdown approaches the 100-V level. A nearly zero temperature coefficient is practical if the FET is biased properly. Most important, the FET constant-current source is usually a very simple, easy-to-design circuit. Either a fixed or an adjustable current source can be constructed with nothing more than a FET, a resistor (fixed or variable), and a battery.

As Figure 2-50 illustrates, the current source can be fixed, variable, or voltage controlled. In Fig. 2-50a the current that flows is I_{DSS}, and in Fig. 2-50b the current is any value below I_{DSS}. In Fig. 2-50c the current range is I_{DSS} down to lower values, since the FET can be biased below I_{DSS}. The basic circuits shown in Fig. 2-50 can be put in series for high voltages, provided the circuits are shunted with balancing resistors. The circuits can also be put in parallel for higher currents. The FETs can be connected in series-opposing for bilateral current limiting.

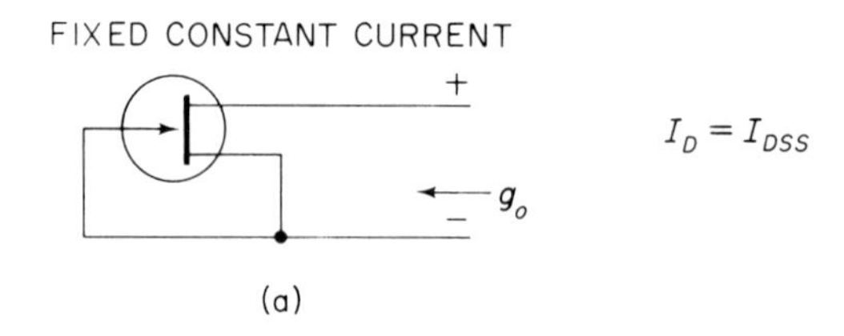

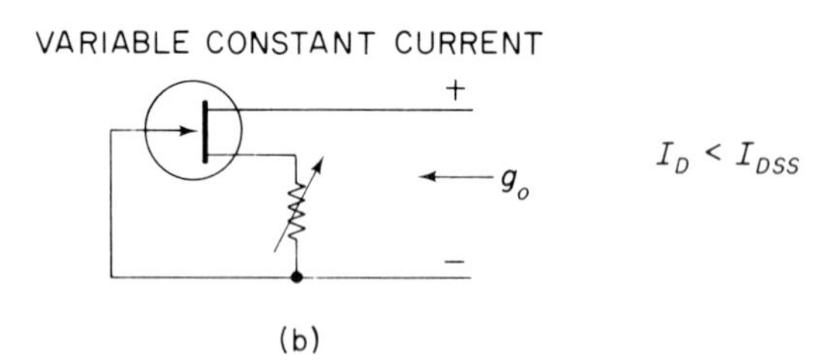

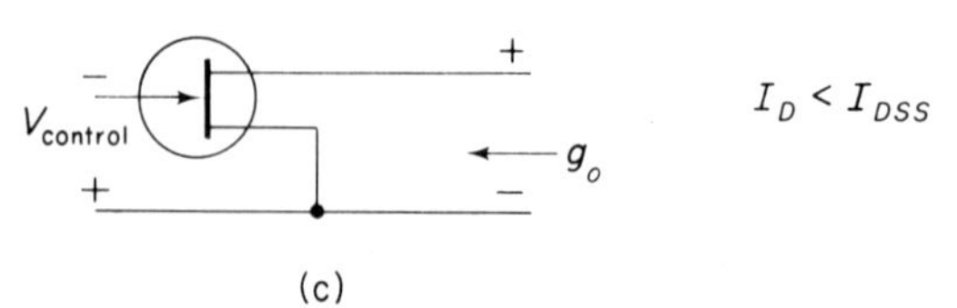

Fig. 2-50 Basic FET constant-current regulators (Courtesy of Motorola Inc., Semiconductor Products Division)

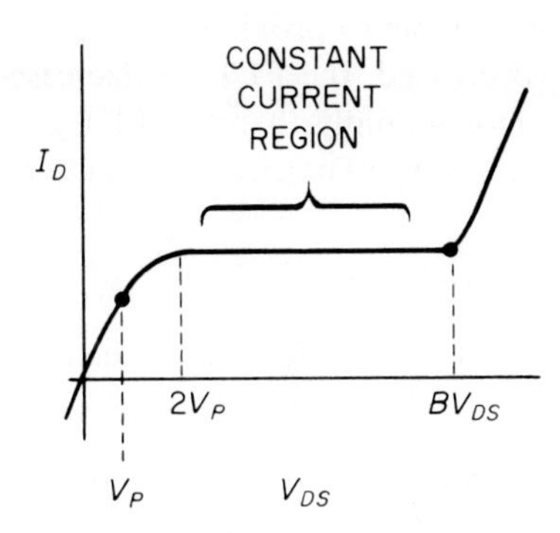

Fig. 2-51 FET output characteristics as constant-current regulator (Courtesy of Motorola Inc., Semiconductor Products Division)

2-20-1 *Fixed Constant Current*

The simplest constant-current circuit is that of Fig. 2-50a. The circuit is essentially a FET with the gate and source shorted. In this case, the FET operates at I_{DSS}, the zero-bias drain current. The circuit output conductance g_0 is equal to the g_{os} of the FET. At low frequencies, g_{os} (the real part of y_{os}) is equal to y_{os}. For higher frequencies, depending upon the value of the FET output capacitance, it may be necessary to add a series inductance to offset the change in y_{os} (or the difference between g_{os} and y_{os}).

The output current I_o of the circuit changes, if V_{DS} changes, according to the relationship

$$\Delta I_o = \Delta V_{DS} g_{os}$$

For example, assume that $g_{os} = y_{os} = 75$ μmhos, and there is a 1-V change in V_{DS}. This will produce a 75-μA change in I_o.

The circuit of Fig. 2-50a will deliver a relatively constant current from about $2V_P$ (the pinch-off or threshold voltage) to BV_{DS} (the breakdown voltage drain-to-source). This range is illustrated in Fig. 2-51.

2-20-2 *Adjustable Constant Current*

With the addition of a source resistor as shown in Fig. 2-50b, the circuit becomes capable of supplying any current below I_{DSS}. The approximate value of gate-source voltage V_{GS} required for a given operating current is

$$V_{GS} = V_P\left(1 - \sqrt{\frac{I_o}{I_{DSS}}}\right)$$

With V_{GS} established, the value of source resistance R_S is

$$R_S = \frac{V_{GS}}{I_o}$$

Resistor R_S can be variable to provide an adjustable current source. As R_S is increased and I_o decreased, the FET g_{os} decreases. The circuit output conductance decreases more rapidly than the FET g_{os} because of the feedback action produced across R_S. The circuit output conductance is

$$g_o = \frac{g_{os}}{1 + R_S(g_{os} + g_{fs})} \approx \frac{g_{os}}{1 + R_S g_{fs}}$$

where g_{fs} is the real part of y_{fs}, the forward transfer admittance.

2-20-3 *Voltage-Controlled Constant Current*

The constant-current circuit can be controlled by an external voltage, as shown in Fig. 2-50c. The approximate value of control voltage (V_{control}) required for a given I_o is

$$V_{\text{control}} \approx V_P\left(1 - \sqrt{\frac{I_o}{I_{DSS}}}\right).$$

For example, assume that V_P is 3 V, I_{DSS} is 1 mA, and the desired I_o is 0.5 mA (I_o must be less than I_{DSS}). Then:

$$V_{\text{control}} \approx 3\left(1 - \sqrt{\frac{0.5}{1}}\right) \approx 0.9 \text{ V}$$

2-20-4 *Cascaded* FET *Current Regulators*

If two FETs are cascaded as shown in Fig. 2-52, a much lower output conductance g_o value (for a given I_o) can be obtained. Here I_o is regulated by Q_1. Note that V_{DS1} is equal to V_{GS2}. The d-c value of I_o is controlled by R_S and Q_1. However, Q_1 and Q_2 both affect current stability: Where $R_S = 0$

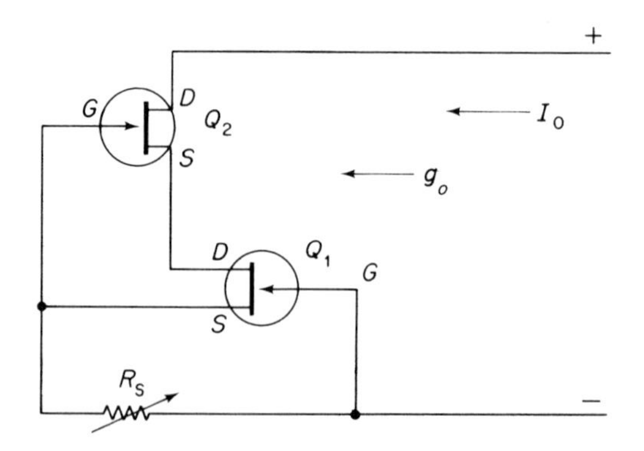

Fig. 2-52 Cascaded FET constant-current source (Courtesy of Motorola Inc., Semiconductor Products Division)

$$I_o = \frac{V_{DS2}\, g_{os1}\, g_{os2}}{g_{os1} + g_{fs2}}$$

and
$$g_o = \frac{g_{os1}\, g_{os2}}{g_{os1} + g_{os2} + g_{fs2}}$$

if $R_S = 0$ and $g_{os1} \approx g_{os2}$,

$$g_o = \frac{g_{os}}{2g_{os} + g_{fs} + R_S(g_{fs2} + g_{os}g_{fs} + g_{os2})} \approx \frac{(g_{os})^2}{g_{fs}(1 + R_S g_{fs})}$$

When designing cascaded FET current sources, care must be exercised to ensure that both FETs are operating with adequate drain-source voltage, preferably $V_{DS} > 2V_P$, and that Q_2 has significantly higher I_{DSS} than Q_1.

2-20-5 Zero-Temperature-Coefficient Current Regulators

The FET current regulators described thus far can be operated as zero-temperature-coefficient (0TC) devices. Of course, the FET must be operated at a specific current I_{DZ}, as discussed in Sec. 2-4. The approximate value of 0TC current I_{DZ} is found by

$$I_{DZ} \approx I_{DSS}\left(\frac{0.63}{V_P}\right)^2$$

The gate-source bias voltage for 0TC is

$$V_{GSZ} \approx V_P - 0.63$$

Note that I_{DZ} increases as I_{DSS} increases. Typically, I_{DZ} can be as high as 1 mA for I_{DSS} units of 20 mA.

By operating the I_D below, but near, I_{DZ}, the temperature coefficient is positive. Conversely, negative temperature coefficients will result if $I_D > I_{DZ}$.

2-20-6 Current Regulators as Voltage References

Low-voltage references can be made by using a FET current source in series with a resistor, as shown in Fig. 2-53. The source simply drives a re-

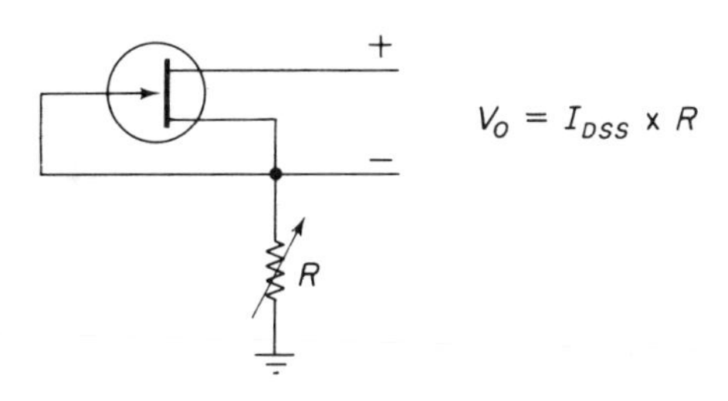

Fig. 2-53 FET current regulator as voltage reference (Courtesy of Motorola Inc., Semiconductor Products Division)

sistor of known value, producing an output reference voltage, which can be determined by Ohm's law. For example, if I_{DSS} is 1 mA and R is 1 kΩ, the voltage reference is 1 V.

2-20-7 FET *Current Regulators as Zener Source Elements*

A FET current regulator can be used in place of the conventional series resistor as a Zener source elment. Such an arrangement has several advantages. For example, the maximum permissible V_{in} is determined by the maximum breakdown voltage of the FET current regulator, rather than the maximum Zener dissipation. Thus, a low-voltage Zener can be used in a high-voltage system (provided the FET can stand the high voltage). Variations in V_{in} have little or no effect on V_{out}. Thus, power efficiency is increased.

Figure 2-54 shows the comparison between a fixed resistor and a FET as a Zener source element.

2-20-8 *Zero-Temperature-Coefficient* FET/*Zener Combinations*

Combining the 0TC feature of a FET regulator with a 0TC Zener diode provides a highly temperature-stable, yet simple, reference voltage. This requires a circuit similar to that of Fig. 2-50b for the FET. However, the source resistor value must be such that the resultant V_{GS} is V_{GSZ}. Refer to Secs. 2-4 and 2-19-5.

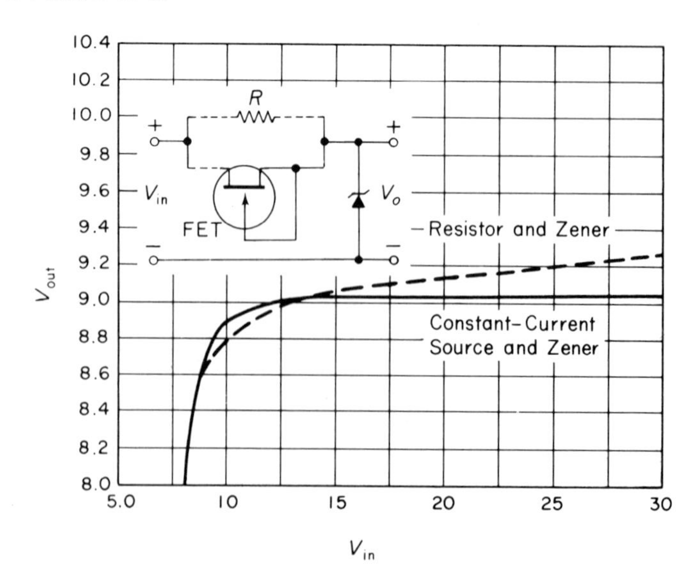

Fig. 2-54 FET current regulator as Zener source element (Courtesy of Motorola Inc., Semiconductor Products Division)

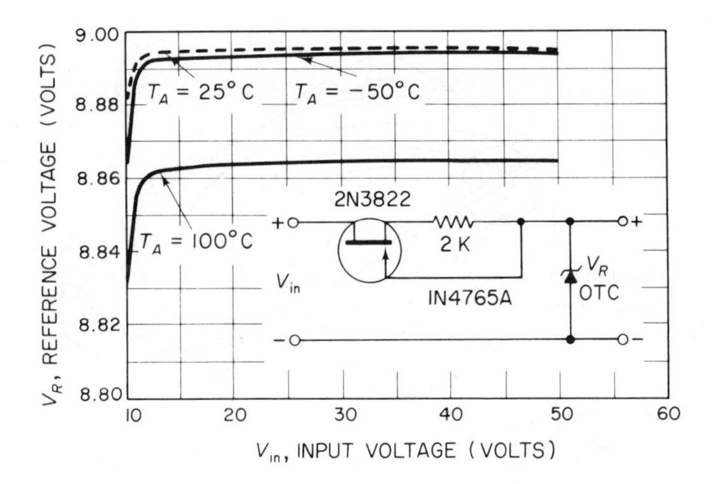

Fig. 2-55 FET current regulator as 0TC voltage reference (Courtesy of Motorola Inc., Semiconductor Products Division)

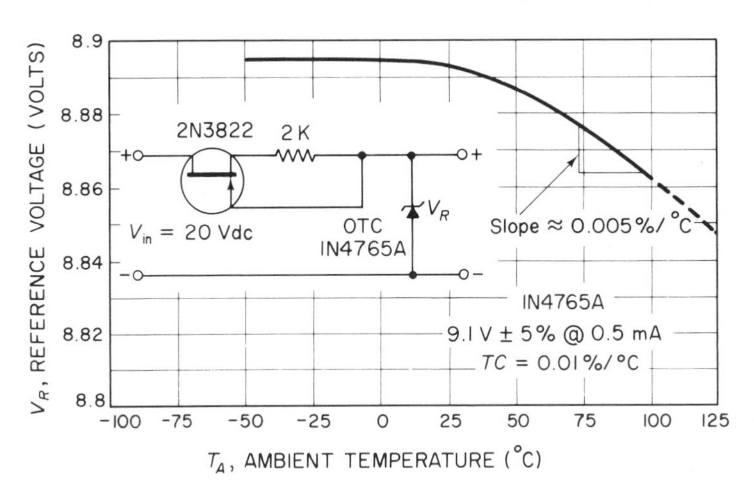

Fig. 2-56 FET current regulator as 0TC voltage reference, across a given temperature range (Courtesy of Motorola Inc., Semiconductor Products Division)

Figures 2-55 and 2-56 show the basic FET/Zener circuit for 0TC operation, as well as the advantages for such an arrangement. Note that the Zener diode *requires a current identical* to the I_{DZ} of the selected FET.

2-21 FET Current-Regulating Diode

The natural development of the FET current regulator (Sec. 2-20, Fig. 2-50) is the Motorola current-regulating "diode" (CRD). In essence, the CRD is an *N*-channel JFET with an *internal* gate-source short. The design of these

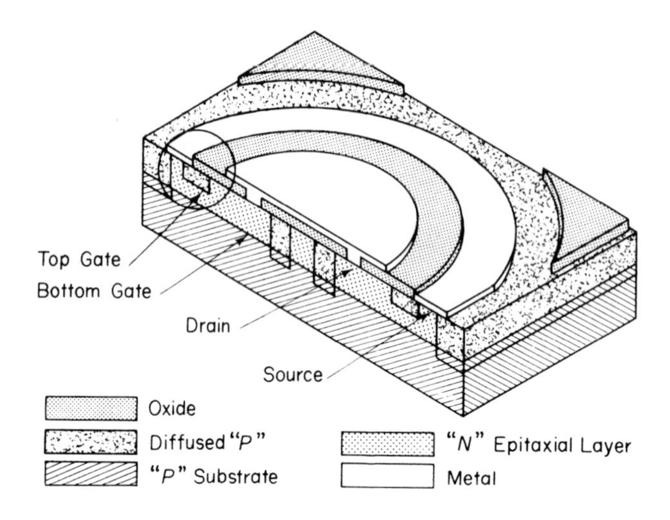

Fig. 2-57 Field effect diode (CRD) construction (Courtesy of Motorola Inc., Semiconductor Products Division)

diodes has been especially optimized for high impedance and current-regulating capability.

A semiconductor diode is defined as a two-electrode semiconductor device. Since the FET CRD meets this definition, it is dubbed a diode. Figure 2-57 shows a typical cross section of the diode. The circled area shows that the metal spans the gate-source areas. Typical CRD current-regulator characteristics, symbols, and definitions are shown in Fig. 2-58.

As shown in Fig. 2-59, the equivalent circuit of the CRD is a current generator in series with a parallel combination of Z_T and a capacitance C_T, or the same generator shunted by a conductance g_T and the same capacitance C_T. The shunt capacitance associated with Motorola's CRDs is about 6 to 8 pF within the useful voltage range of the devices, and is relatively constant. Capacitance does tend to increase and peak as the applied voltage nears V_L, but falls to zero as the voltage goes below V_L.

Figure 2-60 shows performance of several selected CRDs. The minimum operating voltage V_L ranges from 1 to 3 V, while the forward breakdown voltage is specified as 100 V. This gives a 97- to 99-V range of constant-current operation. Futher detailed information is found on the CRD data sheets. The CRDs operate over a $-55°C$ to $+200°C$ range, and have a moderate temperature coefficient (unless a 0TC CRD is selected). Maximum diode dissipation is 600 mW, derated at 4.8 mW/°C above 75°C T_L (lead temperature), as shown in Fig. 2-61.

2-21-1 *Variations in* FET CRD *Configurations*

Various CRD configurations can be implemented to extend their maximum operating voltage and current ranges.

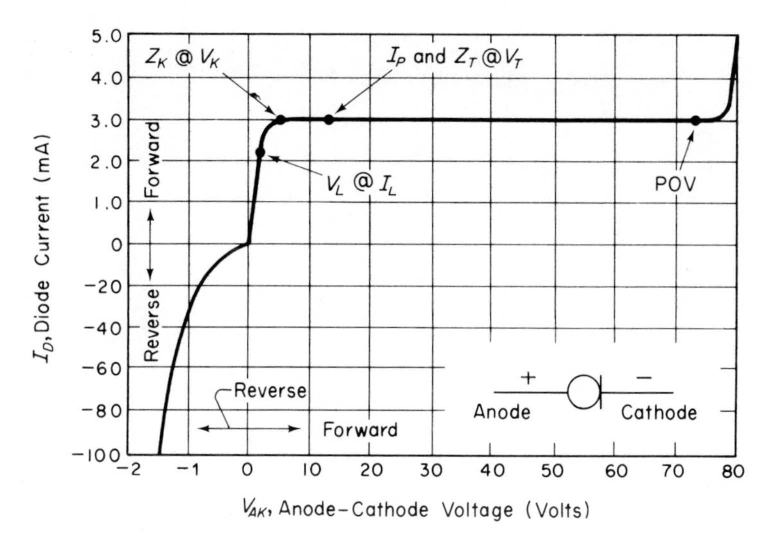

SYMBOLS AND DEFINITIONS

I_D — Diode Current.

I_L — Limiting Current: 80% of I_P minimum used to determine Limiting voltage, V_L.

I_P — Pinch–off Current: Regulator current at specified Test Voltage, V_T.

POV — Peak Operating Voltage: Maximum voltage to be applied to device.

θ_1 — Current Temperature Coefficient.

V_{AK} — Anode-to-cathode Voltage.

V_K — Knee Impedance Test Voltage: Specified voltage used to establish Knee Impedance, Z_K.

V_L — Limiting Voltage: Measured at I_L. V_L, together with Knee AC Impedance, Z_K, indicates the Knee characteristics of the device.

V_T — Test Voltage: Voltage at which I_P and Z_T are specified.

Z_K — Knee AC Impedance at Test Voltage: To test for Z_K, a 90 Hz signal, v_K with RMS value equal to 10% of test voltage, V_K, is superimposed on V_K:

$$Z_K = v_K / i_K$$

where i_K is the resultant ac current due to v_K
To provide the most constant current from the diode, Z_K should be as high as possible; therefore, a minimum value of Z_K is specified.

Z_T — AC Impedance at Test Voltage: Specified as a minimum value. To test for Z_T, a 90 Hz signal with RMS value equal to 10% of Test Voltage, V_T, is superimposed on V_T.

Fig. 2-58 Typical CRD characteristics (Courtesy of Motorola Inc., Semiconductor Products Division)

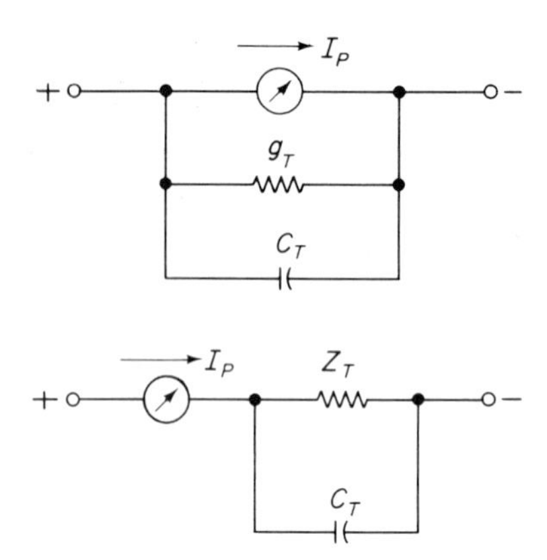

Fig. 2-59 CRD equivalent circuits (Courtesy of Motorola Inc., Semiconductor Products Division)

An extension of the dynamic voltage range is achieved by placing the CRDs in series, as shown in Fig. 2-62a. Here it is necessary to introduce voltage-balancing resistances. The resistive values should be high since they shunt the output resistance. This technique is identical to the current-balancing required when paralleling Zener diodes.

The current range of the diodes can be extended by paralleling CRDs, as shown in Fig. 2-62b. No special precaution is required. The resultant current is the summation of the individual currents. For example, if two 5-mA CRDs are in parallel, the total current is 10 mA.

There are instances where a bipolar circuit (one which regulates or limits in both directions) is useful. In this case, the CRDs are connected in series-opposing fashion, as shown in Fig. 2-62c. During the cycle when one CRD is limiting, the other is a forward-biased junction, producing a diode voltage drop.

2-21-2 General Applications of CRDs

Low-voltage reference. As shown in Fig. 2-63, the CRD can be connected in series with a resistor to form a low-voltage reference source. Such a circuit is particularly effective at voltages below 2.5 V (below normal Zener values), and can be used as a voltage reference up to about 6 V. CRDs serve well as a precision millivolt reference source. The CRD drives a known resistance value, producing an output reference voltage, the value of which is determined by Ohm's law.

Zener diode source element. As shown in Fig. 2-64, the CRD can be used in place of the series resistor in a Zener supply. Figures 2-64a and 2-64b illustrate the conventional series resistor and CRD-Zener regulator circuits, respectively.

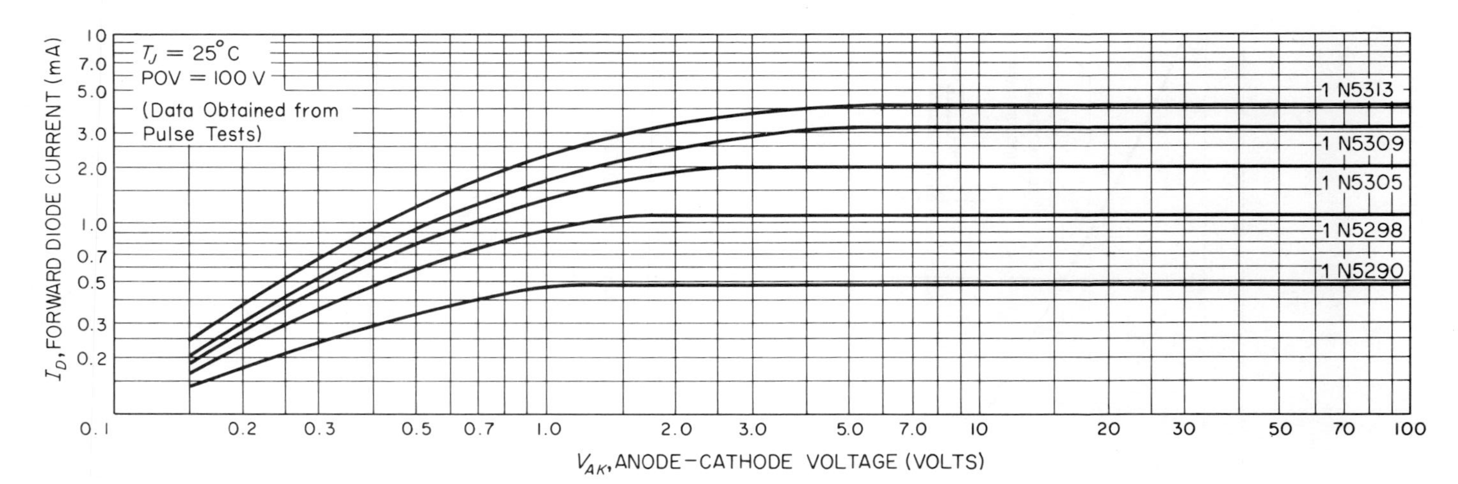

Fig. 2-60 Typical CRD forward characteristics (Courtesy of Motorola Inc., Semiconductor Products Division)

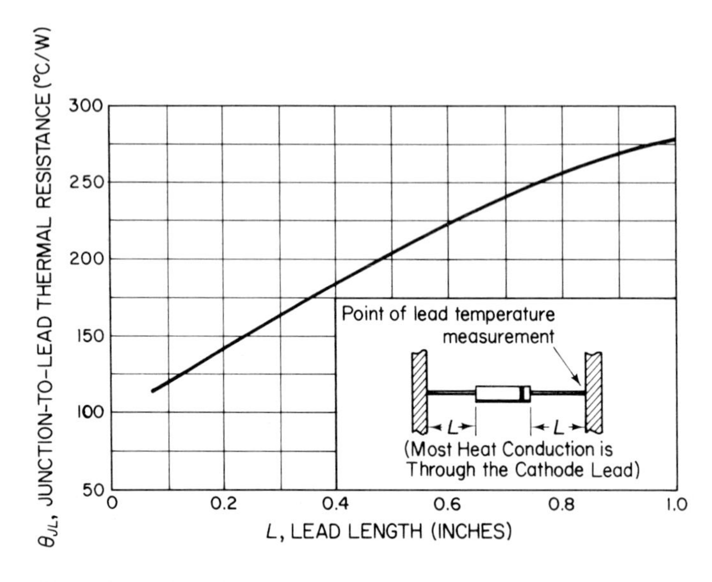

Fig. 2-61 Typical CRD thermal resistance characteristics (Courtesy of Motorola Inc., Semiconductor Products Division)

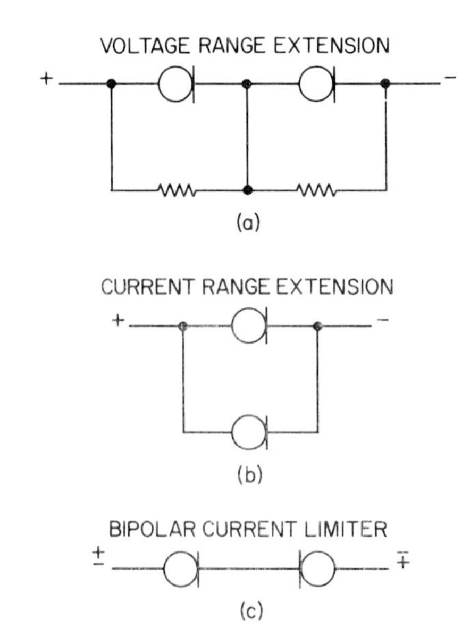

Fig. 2-62 Variations in CRD configurations (Courtesy of Motorola Inc., Semiconductor Products Division)

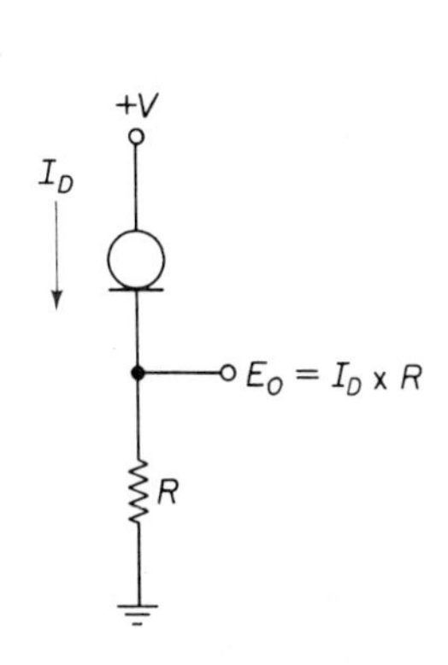

Fig. 2-63 CRD as low-voltage regulator (Courtesy of Motorola Inc., Semiconductor Products Division)

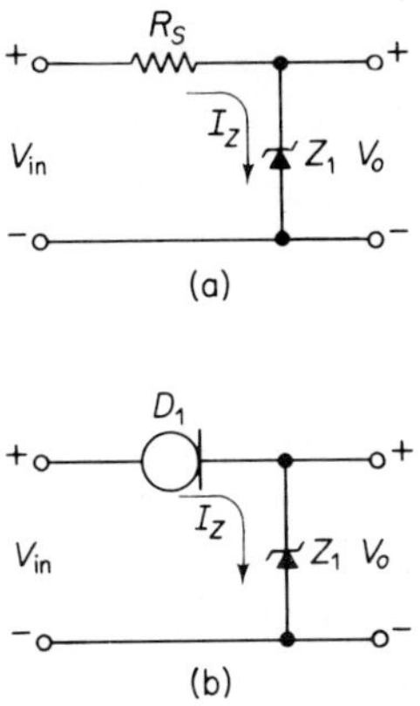

Fig. 2-64 CRD as Zener diode source element (Courtesy of Motorola Inc., Semiconductor Products Division)

Distinct advantages of the Zener-CRD combination are that the maximum permissible V_{in} is determined by the maximum CRD voltage (100 V), rather than the maximum Zener dissipation, and that variations in V_{in} have little effect on V_{out}.

In the circuit of Fig. 2-64a,

$$\Delta V_{out} \approx \Delta V_{in} \frac{Z_{ZT}}{R_S + Z_{ZT}}$$

In the circuit of Fig. 2-64b,

$$\Delta V_{out} \approx \Delta V_{in} \frac{Z_{ZT}}{Z_T + Z_{ZT}}$$

where R_S is the series resistance, Z_T is the dynamic impedance of the CRD, and Z_{ZT} is the Zener dynamic impedance.

Since Z_{ZT} is relatively low, it is obvious that as R_S increases in value, the regulation, V_{out}/V_{in}, is improved. In practical circuits, R_S is normally in the hundreds or thousands of ohms. A typical CRD can present a Z_T of many megohms. This produces three to four times better regulation.

As V_{in} varies, I_Z also varies according to

$$\Delta I_Z = \frac{\Delta V_{in}}{R_S + Z_{ZT}}$$

From the relationship given, it is obvious that as V_{in} increases, the current through and, consequently, the power dissipated in the Zener increases. V_{in} is thus limited by the power rating of the Zener, when R_S is used.

A most absolute voltage reference can be impelemented by using a Zener-CRD combination, with zero temperature coefficient, as shown in Fig. 2-65.

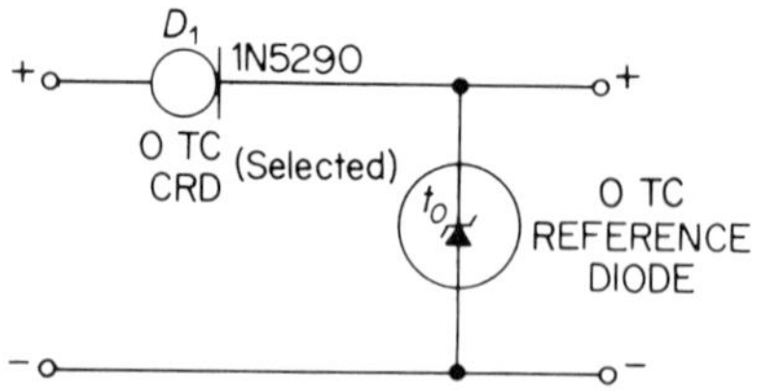

Fig. 2-65 0TC CRD used with 0TC Zener to form most absolute voltage reference (Courtesy of Motorola Inc., Semiconductor Products Division)

The CRD shows 0TC at currents around 0.5 mA. A Motorola family of 0TC Zener diodes also operates at currents of 0.5 mA. A combination of the two with 0TC characteristics will yield a constant reference voltage between the input voltage limits of $2V_L + V_Z$ to $POV + V_Z$ (in practical terms, about 8.4 to 106.4 V with a 6.4-V Zener). The temperature range of such a configuration is typically about 0°C to 100°C. A change in V_{out} of about 10 mV can be expected over the V_{in} range. A temperature change of 0.001 %/°C can be expected over the temperature range.

When voltages lower than system power supplies are required and a Zener is used, another advantage of the CRD is *decoupling of noise or ripple on the supply lines*. Due to the high ratio of the dynamic impedance of the CRD, as compared to the Zener, an attenuation of about 100 dB can be realized at frequencies up to several hundred kHz.

D-C coupling. As shown in Fig. 2-66, the CRD can be used effectively in coupling circuits between transistors. A standard form of d-c coupling is shown in Fig. 2-66a. The circuit is much improved by the substitution of a Zener diode for R_1, as shown in Fig. 2-66b. This modification substantially reduces the loss of gain introduced by the voltage division of R_1/R_2, since the resistance of the Zener is a fraction of a practical R_1 resistance value. A similar increase in gain is achieved by the substitution of a CRD for R_2 (Fig. 2-66c) since the resistance presented by the CRD is many times that of a practical R_2 resistance value. When R_1 and R_2 are replaced by a Zener and a CRD, respectively (Fig. 2-66b), the maximum voltage transfer occurs. In this arrangement, there is practically no signal voltage division, since the Zener impedance is low and the CRD impedance is high.

Constant-current source for differential amplifiers. The common-mode rejection ratio of emitter-coupled differential amplifiers is directly proportional to the common-emitter impedance. For this reason, current sourcing in the emitter is commonly used. Figure 2-67a shows a resistor-voltage combination, while Fig. 2-67b is the classic biased-transistor version. An improvement over both of these methods is obtained by using a CRD, as shown in Fig. 2-67c.

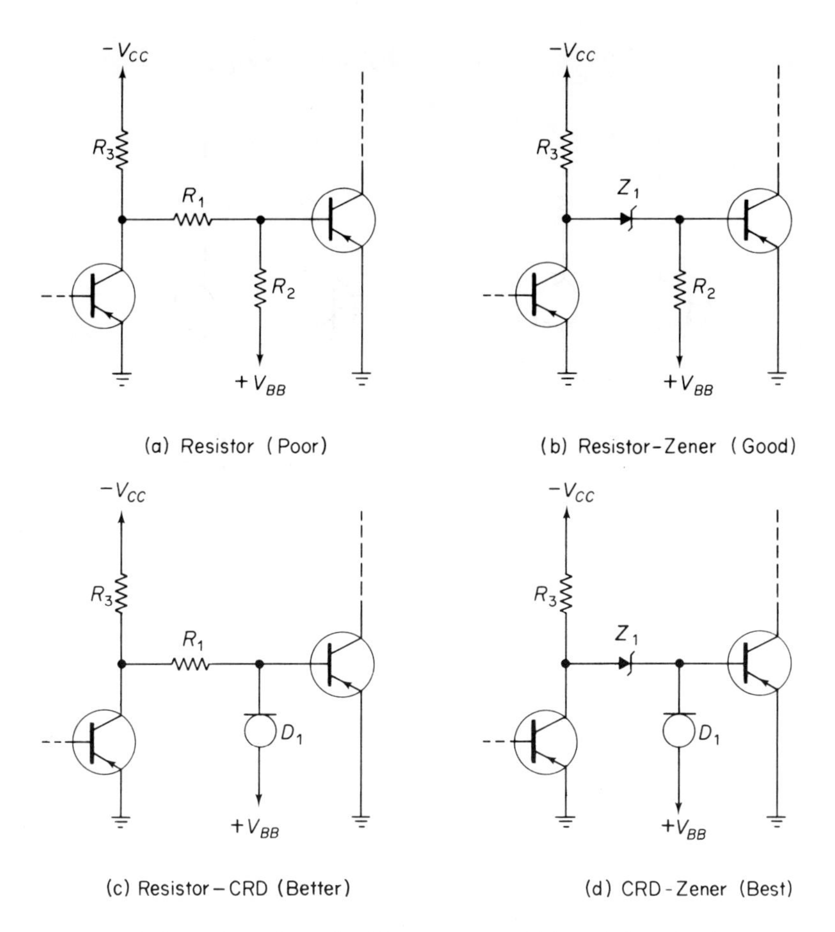

Fig. 2-66 D-C coupling with CRD (Courtesy of Motorola Inc., Semiconductor Products Division)

When high input impedance is required, the differential amplifier uses the Darlington input configuration of Fig. 2-68. The modified amplifier of Fig. 2-68b has three current sources. Current source D_2 provides the high common-mode rejection, while D_1 and D_2 provide fixed currents in Q_1 and Q_4. Using this method, the characteristics of transistors Q_1 and Q_4 are not a function of the betas of the input transistors Q_2 and Q_3, as will be the case with a standard Darlington circuit.

Emitter-follower resistor. The emitter resistor of emitter-followers can often be replaced by a CRD, resulting in a significant increase in input impedance, a gain closer to unity, and a lower transistor dissipation when supplying a heavy external load. Figure 2-69 illustrates an example where an

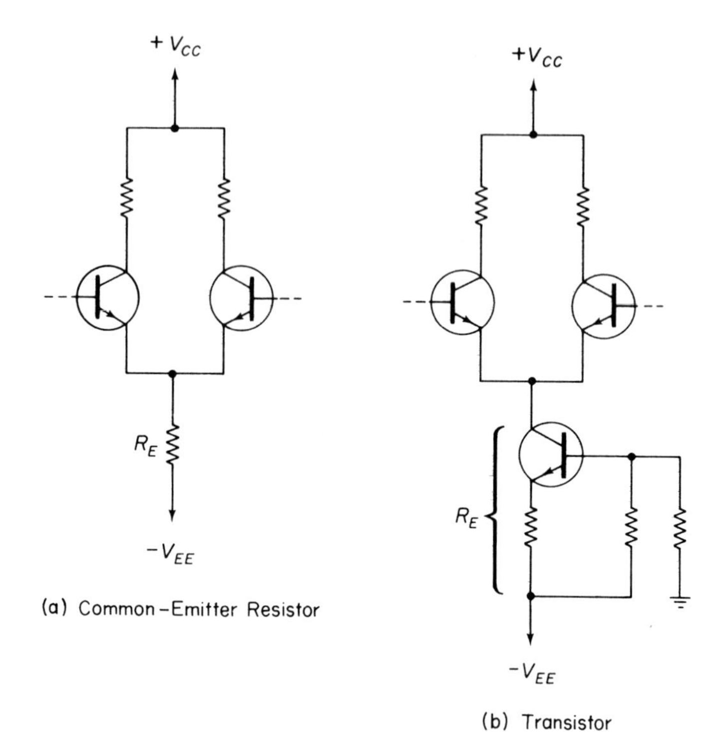

(a) Common-Emitter Resistor

(b) Transistor

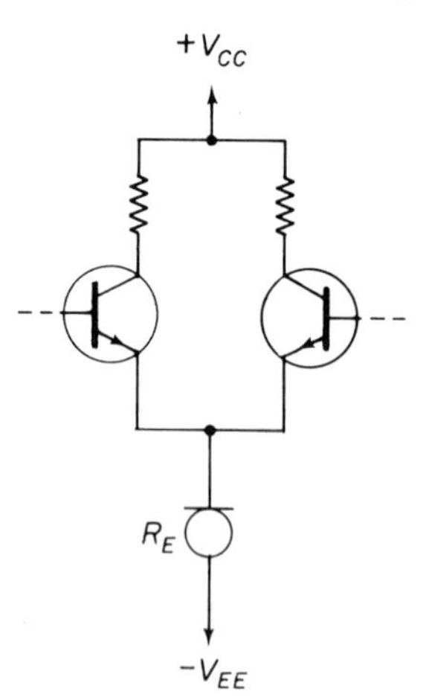

(c) Current-Regulator Diode

Fig. 2-67 CRD as current source for differential amplifier (Courtesy of Motorola Inc., Semiconductor Products Division)

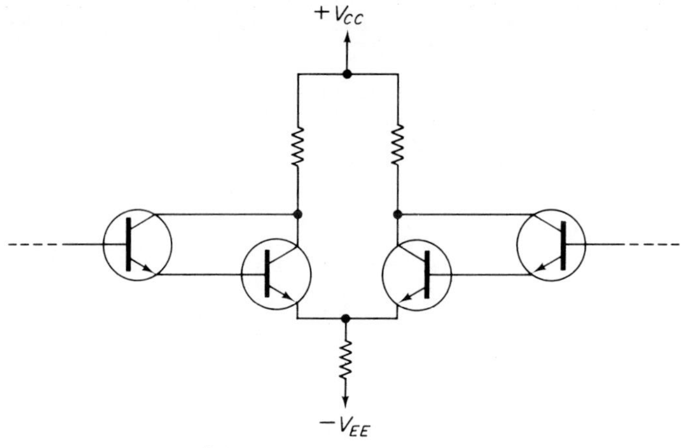

(a) Common-Emitter Resistor

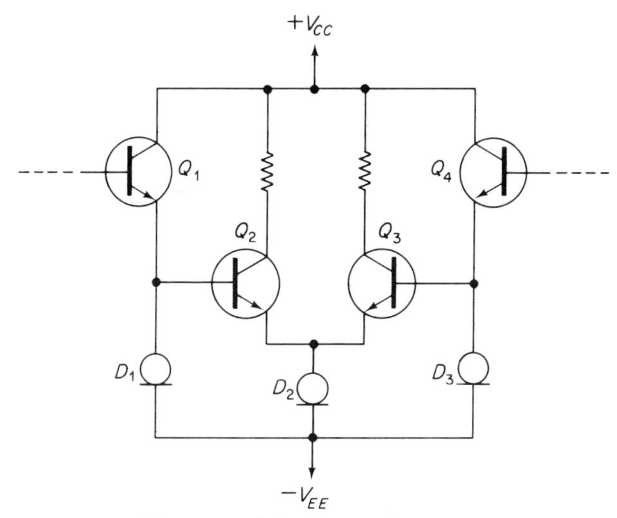

(b) Improved Emitter Current Source

Fig. 2-68 CRD as emitter current source for differential amplifiers (Courtesy of Motorola Inc., Semiconductor Products Division)

output of 5-V peak is required in a 600-ohm load, with ±10-V supplies. When the input peak is +5 V, the load current is 8.33 mA. The drop across R_1 is 5 V and, in order that Q_1 is not cut off, R_1 must pass more than 8.33 mA. Thus, R_1 should be less than 600 ohms. Under quiescent conditions, the emitter of Q_1 is near zero. Quiescent current is 16.7 mA, and Q_1 dissipates 167 mW. When R_1 is replaced by an 8.33 mA CRD, the quiescent dissipation

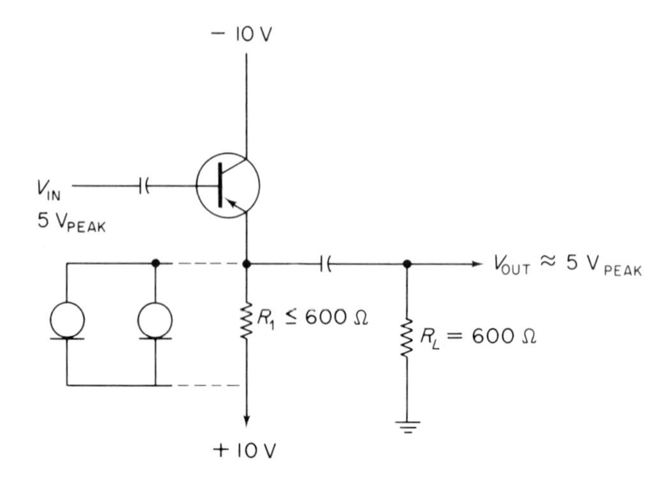

Fig. 2-69 CRD as emitter-follower resistor (Courtesy of Motorola Inc., Semiconductor Products Division)

is 83 mW. Thus, power consumption is cut in half. Since the transistor load is 600 ohms instead of 300, the voltage gain is nearer unity. Note that the parallel operation of two CRDs is used to obtain the 8.33 mA.

Collector load resistor. In amplifier circuits where the collector voltage is defined by an external feedback loop, the collector load resistor can be replaced by a CRD to give a greatly increased voltage gain. Using T equivalent parameters, the voltage gain A_V of a grounded-emitter amplifier approaches the value given by

$$A_V = \frac{t_c}{t_e}$$

In order to relaize this gain, a large collector load is required (the higher the value of t_c in relation to t_e, the higher the gain). This can be obtained by replacing the collector resistor with a CRD. Such an arrangement can produce voltage gains from 700 to 1000 with currently available small-signal transistors. Of course, voltage gain will be reduced if the amplifier is substantially loaded.

Similar results can be otained by using the CRD as the drain element in a FET amplifier. Such a circuit is shown in Fig. 2-70. For best results, the FET and CRD should be selected so that they have substantially the same current characteristics. Source resistor R_S is used to balance the drain currents and to adjust for maximum voltage gain. If the source resistor must be shorted out, or raised to some very large value, this is an indication that the FET and CRD are not properly matched. As shown in Fig. 2-70, a FET can be used in place of the CRD. The FET must be connected as a current regulator, as described in Sec. 2-19. With either circuit, R_S is adjusted for maximum voltage gain.

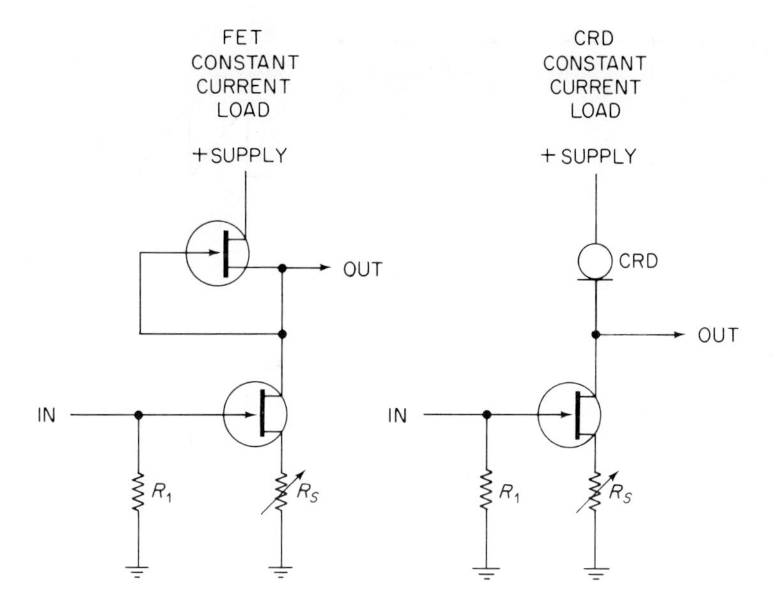

Fig. 2-70 Using a CRD (or FET) as drain element in a FET amplifier

2-21-3 Specific Applications of CRDs

Simple sawtooth generator. The two circuits of Fig. 2-71 make use of the Motorola four-layer diode and CRD to provide simple, fixed-frequency sawtooth generators with linear output waveforms. As shown, the circuits provide both positive- and negative-going ramps.

The principal design equation for these circuits is

$$T = \frac{CV_{BR}}{I_P}$$

where T = period of one cycle
I_P = pinch-off current of the CRD
C = timing capacitor in μF
V_{BS} = breakover voltage of the four-layer diode

Triangular wave generator. The use of the CRD in series-opposing fashion makes possible the generation of a high-quality triangular wave from a sine- or square-wave source, as shown in Fig. 2-72. The square-wave drive results in a better waveform at the zero crossing. Here, the output frequency is

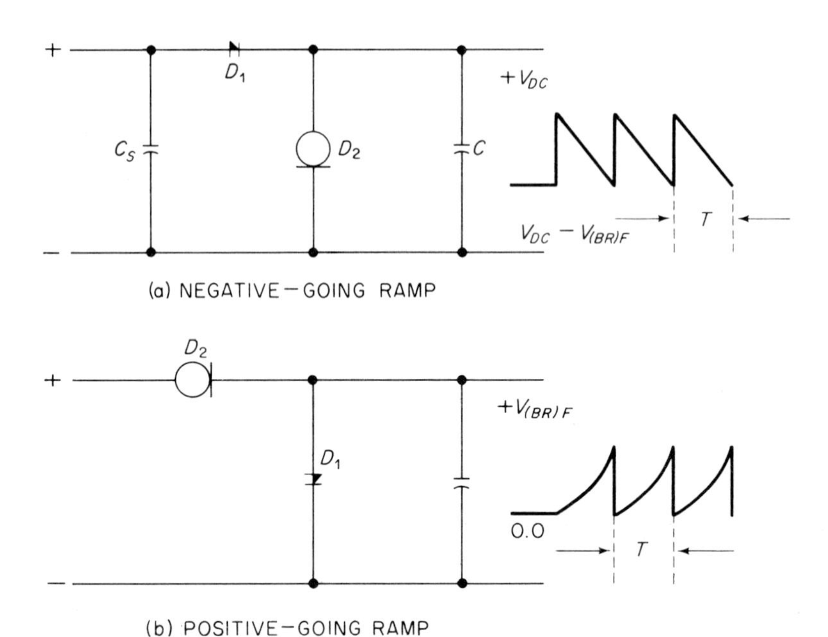

Fig. 2-71 CRD as sawtooth generator (Courtesy of Motorola Inc., Semiconductor Products Division)

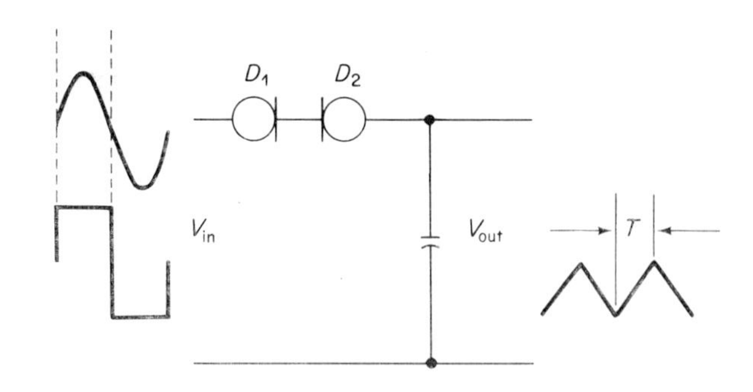

Fig. 2-72 CRD as triangular wave generator (Courtesy of Motorola Inc., Semiconductor Products Division)

identical to the input frequency. The peak-to-peak amplitude is given by

$$V_o(p\text{-}p) = \frac{I_s}{C}$$

where $V_o(p\text{-}p)$ is in volts, C is in μF, I is in mA, and t is in mS.

Square-wave generator or an improved clipper. A popular solid-state circuit, and the improved CRD verison, are shown in Fig. 2-73. In either circuit, the output frequency is the same as the input frequency. The peak

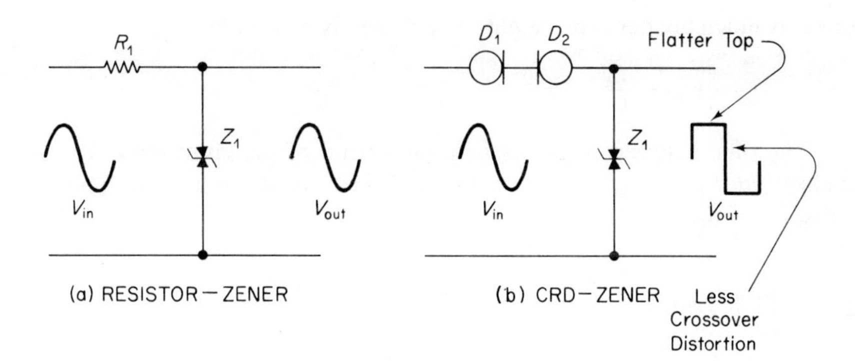

Fig. 2-73 CRD as square-wave generator or clipper (Courtesy of Motorola Inc., Semiconductor Products Division)

value of the output waveform is:

$$V_{o(pk)} \approx (0.7 + V_Z) \text{ volts}$$

where V_Z is the Zener voltage in volts.

The improved CRD circuit has the additional advantages of increased efficiency, and reduced power dissipation in the Zener.

Stairstep generator. The circuit of Fig. 2-74 operates on the same principle as the triangular wave generator. A single CRD is used, and the ratio of

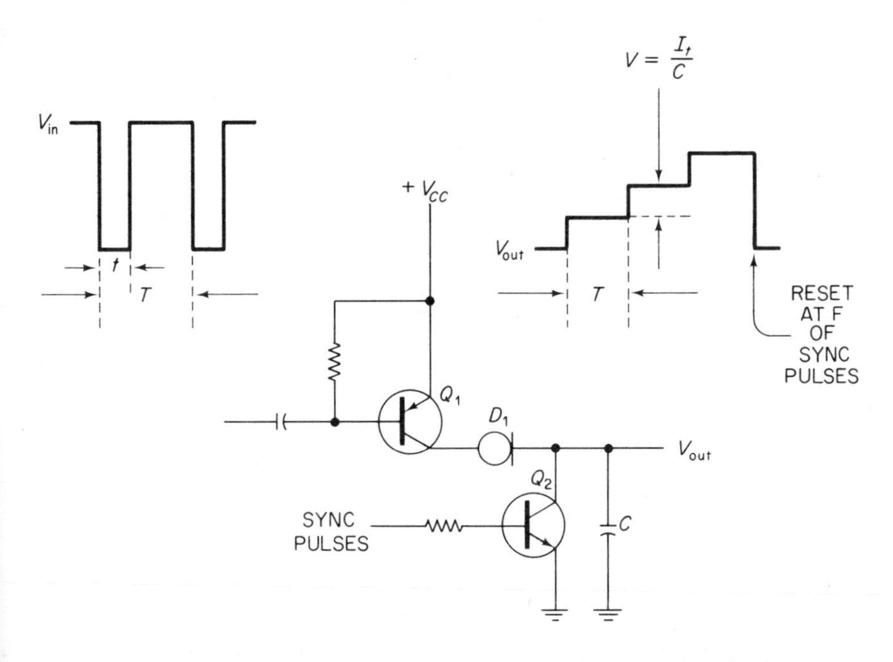

Fig. 2-74 CRD as stairstep generator (Courtesy of Motorola Inc., Semiconductor Products Division)

I to C is much greater. The height of each step is defined by

$$V_{o(p\text{-}p)} \frac{I_s}{C}$$

The time between steps is set by the period of the input pulse. In this application, V_{CC}-$V_{CE(\text{sat})}$ of Q_1 must exceed the level of the highest step by at least $2V_L$ of the CRD.

2-22 FET Voltage-Variable Resistors

When a FET is used in the ohmic region below V_P (Fig. 2-3), the FET is simply a voltage-controlled resistor. The uses of a FET in this region are virtually unlimited. In effect, anywhere a voltage-controlled resistor under small-signal conditions is needed, a FET can be used. Of course, heavy currents cannot be handled by present-day FETs.

Of prime interest is the curve of Fig. 2-75. Here the R_{DS} (normalized to R'_{DS} or the drain-source resistance at zero bias) is plotted against a ratio of V_{GS}/V_P. That is, the drain-source resistance is given as a function of the gate-source voltage V_{GS}, normalized to V_P. Here V_P is defined as the gate-source voltage required to reduce I_D to $0.001 I_{DSS}$. Note that when V_{GS} is equal to V_P, the drain-source resistance increases rapidly (to infinity, in theory).

The curve of Fig. 2-75 is valid for several families of *depletion* IGFETs and JFETs. However, enhancement FETs are excluded.

Dynamic attenuator. Figure 2-76 shows two FETs operated as voltage-variable resistors to form a dynamic attenuator. The curves of Fig. 2-76 show

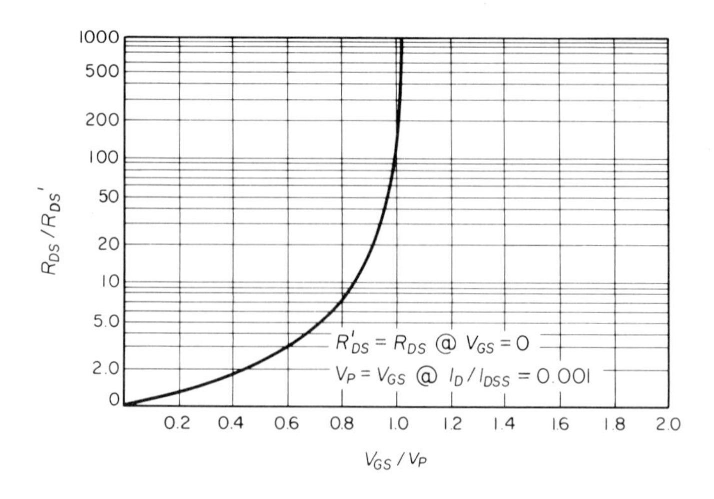

Fig. 2-75 Normalized R_{DS} data for depletion silicon FETs (Courtesy of Motorola Inc., Semiconductor Products Division)

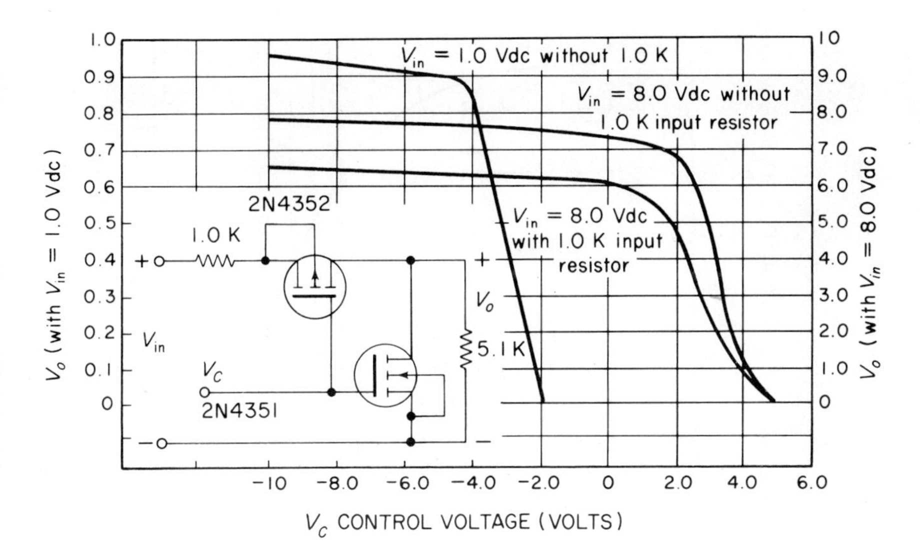

Fig. 2-76 FET as a wide-range dynamic attenuator (Courtesy of Motorola Inc., Semiconductor Products Division)

the ability of a change in V_{GS} to vary R_{DS}. Here a change of only 1 V swings the attenuation over 50 per cent of the range. As the resistance of one FET increases, that of the other decreases. If the drain-source voltage is permitted to exceed V_P, as it has in the case of $V_{in} = 8$ V, it appears that R_{DS} still varies. (Here the series element is acting as a controlled-current source to the load.)

Voltage-controlled filters. When a FET operated as a voltage-controlled resistor is combined with passive components such as capacitors or inductors, voltage-controlled frequency tuning can be implemented. Figure 2-77 shows a voltage-controlled low-pass filter. Here the change in V_{GS} of 1 V changes the upper 3-dB frequency point by a decade.

A high-pass filter can be implemented by transposing the FETs and capacitors (capacitors in series with the line, and FETs across the line). Likewise, high-pass and low-pass filters can be combined to form band-pass filters. Refer to the author's *Handbook of Simplified Solid-State Circuit Design* for a detailed discussion of RC filters. The following is a summary.

Filter attenuation is rated in terms of dB drop at a given frequency. Generally, RC filters are designed to produce a 3-dB drop (to 0.707 of input) at a selected cutoff frequency. The relationships for capacitance and resistance values versus cutoff frequency for RC filters with a 3-dB drop are

$$\text{cutoff frequency} = \frac{1}{6.28RC}$$

where C is in farads, R in ohms, and f in Hz

A single RC filter will provide a gradual transition from the passband to the cutoff region. If a rapid transition is necessary for design, two or more

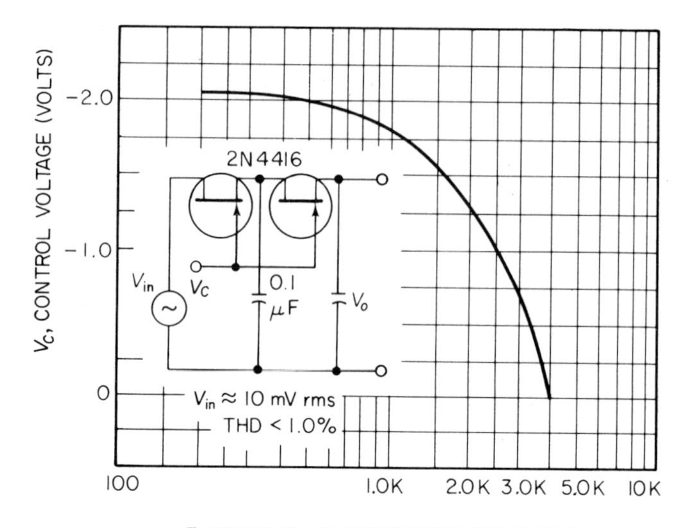

Fig. 2-77 FET as voltage-controlled low-pass filter (Courtesy of Motorola Inc., Semiconductor Products Division)

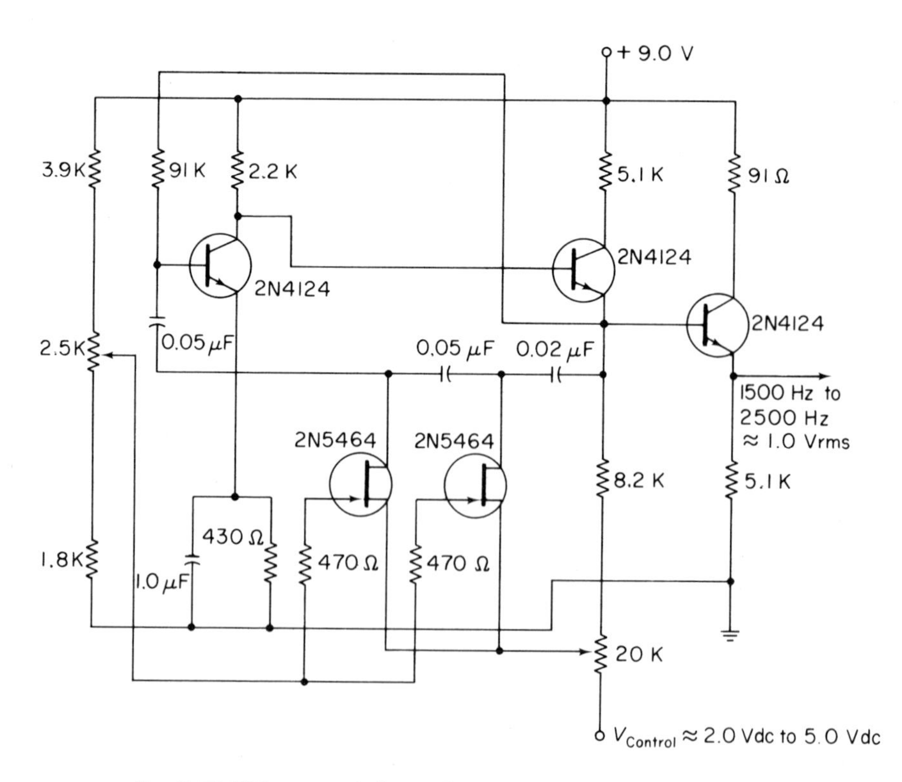

Fig. 2-78 FET as control element in voltage-controlled oscillator (Courtesy of Motorola Inc., Semiconductor Products Division)

RC filter stages can be combined. (Two low-pass stages are shown combined in Fig. 2-77.) As a rule of thumb, each stage will increase the attenuation by 6 dB at the cutoff frequency.

If the drain-source resistance for a given control voltage is known (or can be measured), it is a simple matter to calculate the required value of C to produce a 3-dB drop. For example, assume that the drain-source resistance is 1 kΩ at a given control voltage, and it is desired to produce a 3-dB drop at 300 Hz. The value of C is

$$C = \frac{1}{6.28RF} = \frac{1}{6.28 \times 1000 \times 300} = 0.5\ \mu\text{F}$$

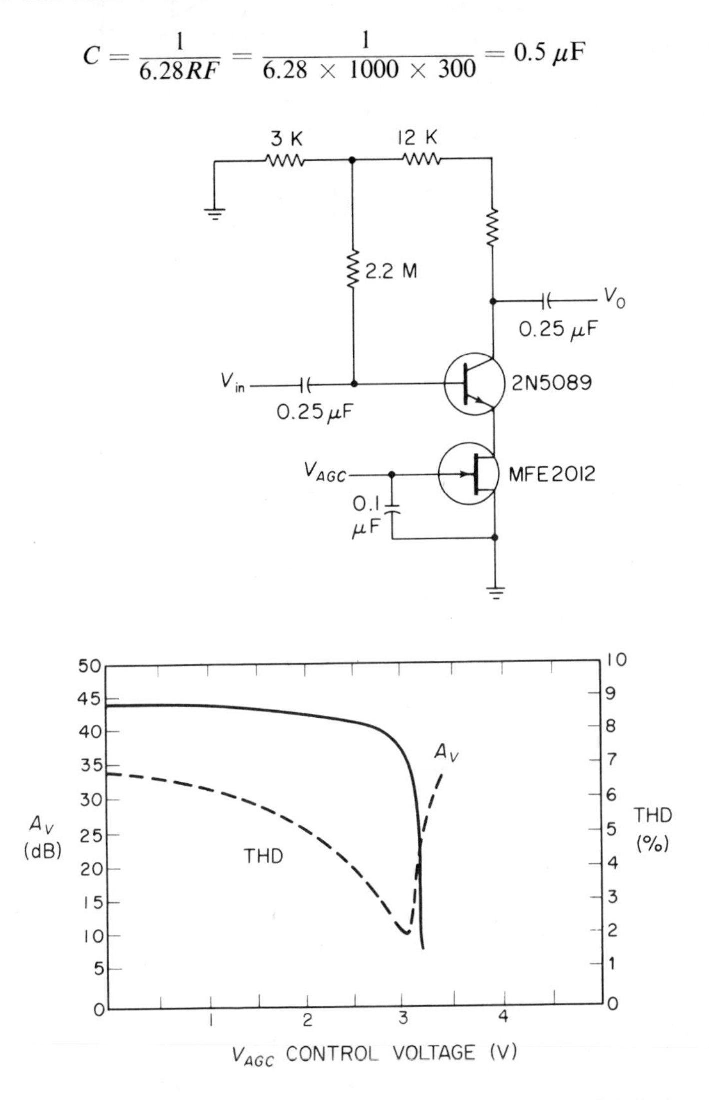

Fig. 2-79 FET operating as a voltage-controlled resistor in an AGC circuit (Courtesy of Motorola Inc., Semiconductor Products Division)

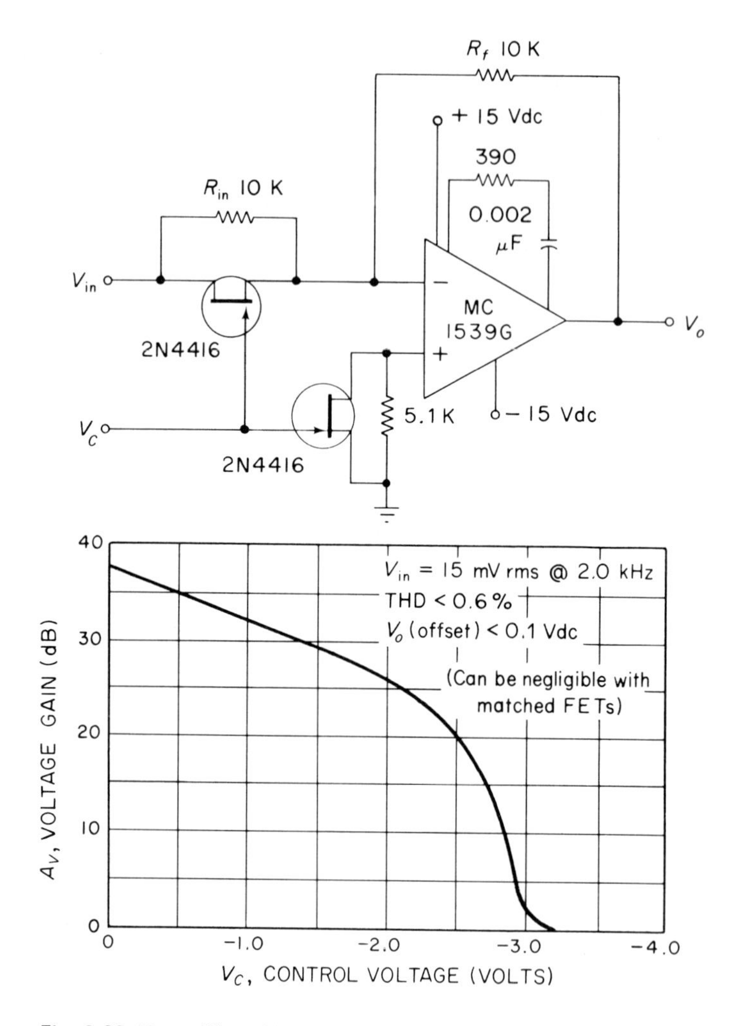

Fig. 2-80 IC amplifier with FET AGC (Courtesy of Motorola Inc., Semiconductor Products Division)

If two identical stages are used, the drop will be approximately 6 dB greater, or 9 dB at 300 Hz. If three stages are used, the drop is 15 dB, and so on.

Voltage-controlled phase-shift oscillator. When a phase-shifting network is included in an amplifier's feedback loop, as shown in Fig. 2-78, a voltage-controlled oscillator is formed. This oscillator produces a sine wave, and is linear over the range of frequency indicated.

Automatic gain control (AGC). A FET operated as a voltage-controlled resistor can be combined with other active devices to form a simple AGC circuit, such as shown in Fig. 2-79. Here the FET changes the voltage gain over a 30-dB range with only a 1-V change in V_{GS}. This circuit has some harmonic distortion, due to the unbypassed emitter degeneration.

When the FET is used in conjunction with an operational amplifier, as shown in Fig. 2-80, the gain can be made variable by changing the ratio of R_f/R_{in}. This is also a rather simple AGC circuit, but with improved distortion characteristics.

Note that a FET is used across the R_{in} resistor, and across the offset-minimizing resistor (at the + input of the *op*-amp). Offset voltage at the output of an *op*-amp is due to input bias currents. These currents are normally minimized by placing a resistance equivalent to the parallel combination of R_f and R_{in} at the noninverting (+) input of the *op*-amp. If an FET is used only across R_{in}, the input impedance will change with variations in control voltage. This will shift (or offset) the output voltage. The FET across the minimizing resistor permits the resistance to vary according to the variation in R_{in}. In turn, this minimizes the change in output offset voltage due to the changing R_{in}. The value of R_{in} should equal R_f. This will insure that gain will not drop below unity.

2-23 FET Switches and Choppers

Field effect transistors make excellent switches and choppers for a multitude of applications such as modulators, demodulators, sample-and-hold systems, mixing, multiplexing or gating, and many more.

Several important characteristics of the FET make it almost ideal for these applications.

One advantage is that there is no inherent offset voltage associated with FETs as there is with bipolar transistors since the conduction path between drain and source is predominantly resistive. In both JFET and IGFET devices, the conduction channel is either depleted or enhanced by controlling an induced field.

The d-c gate input impedance of a FET is also extremely high, and requires little control power. In JFETs, the control signal looks into a reverse-biased diode. For IGFETs, the gate insulation is a high-resistance oxide or nitrite. Therefore, the impedance is determined by the properties of the insulation layer.

Another important advantage of the FET is the exceptionally high ratio of OFF resistance to ON resistance of the drain-source channel. The resistance can be as low as several ohms in the ON condition, and higher than thousands of megohms in the OFF condition.

The main disadvantage of FETs is the capacitance between the gate and drain, and the gate and source. This capacitance feeds through part of the gate control voltage to the signal path. These capacitances are detrimental to high-frequency signal isolation, and also impose a limitation on response times.

2-23-1 FET *Characteristics Applicable to Analog Switches and Choppers*

Figure 2-81 shows the ohmic region of a FET expanded for both positive and negative values of V_{DS}. In the ohmic region, when the FET is fully ON, there is a linear relationship between drain current I_D and drain-to-source voltage V_{DS}. The magnitude of this resistance can be changed by varying the gate-to-source voltage V_{GS}. It is in the ohmic region that the FET is useful for both the chopper and analog switch applications.

As shown in Fig. 2-81, to stay in the ohmic region, the drain current must be kept within narrow limits. In other words, a relatively large value of load resistance R_L is required.

Leakage current $I_{D(\mathrm{OFF})}$ versus temperature for a 3N126 (an *N*-channel JFET) is plotted in Fig. 2-82. A very low value of leakage current is important in chopper applications, since this leakage current appears in the output circuit and produces an error voltage.

$I_{D(\mathrm{OFF})}$ versus temperature for a 2N4352 *P*-channel IGFET is shown in Fig. 2-83. Below 100°C the leakage is so low that accurate readings are more dependent upon available equipment and measurement techniques than on the magnitude of the leakage current. However, the curve of Fig. 2-83 can be projected back to room temperature (shown with dotted lines) and an $I_{D(\mathrm{OFF})}$ of approximately 0.003 pA is read.

At room temperatures, surface and package leakage account for con-

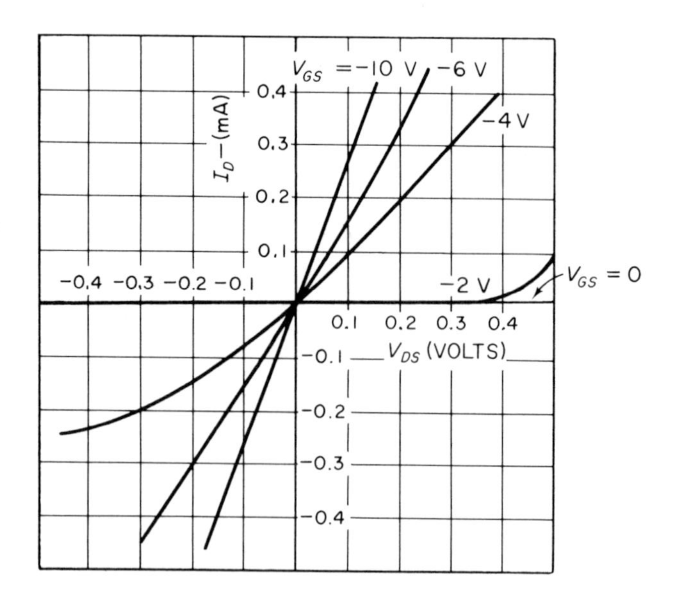

Fig. 2-81 2N4352 low-level (ohmic region) output characteristic (Courtesy of Motorola Inc., Semiconductor Products Division)

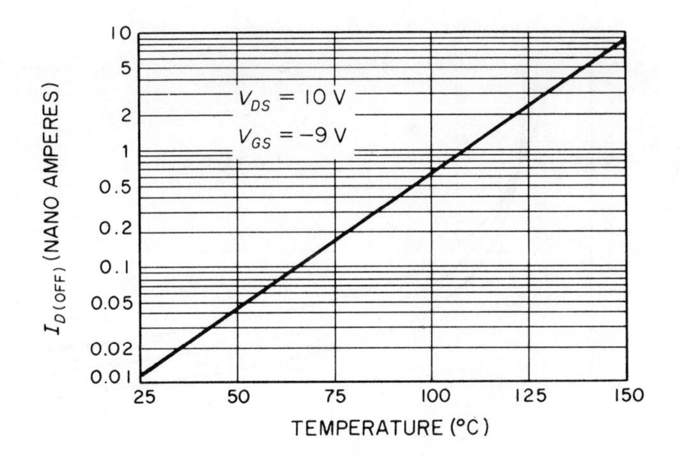

Fig. 2-82 3N126 $I_{D(OFF)}$ versus temperature (Courtesy of Motorola Inc., Semiconductor Products Division)

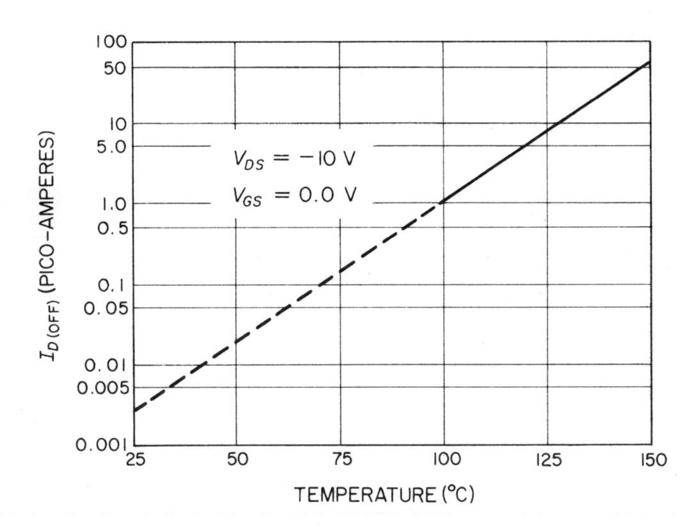

Fig. 2-83 2N4352 $I_{D(OFF)}$ versus temperature (Courtesy of Motorola Inc., Semiconductor Products Division)

siderably more than I_{DSS} with a resulting room-temperature leakage of about 0.5 pA. This low leakage current indicates that the OFF voltage error (caused by leakage) will be negligible for most chopper and analog switching circuits. For an enhancement mode IGFET, $I_{D(\mathrm{OFF})} = I_{DSS}$.

Drain-to-source resistance (R_{DS}), when the FET is ON, is a very important characteristic in both chopper and analog switching circuits. Figure 2-84 shows R_{DS} versus V_{GS} for three values of temperature for a 2N4352.

On a static basis, there is interest in only two states of the FET—fully ON or fully OFF. The 2N4352 (a *P*-channel IGFET) needs a negative potential

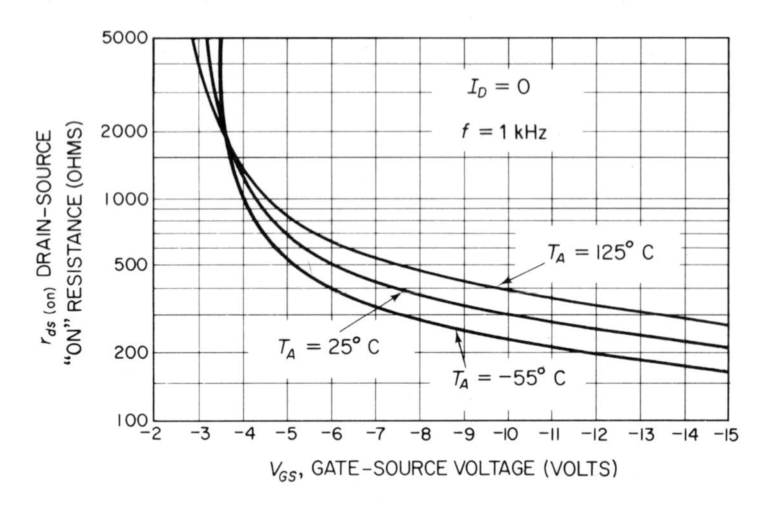

Fig. 2-84 2N4352 drain-source ON resistance (Courtesy of Motorola Inc., Semiconductor Products Division)

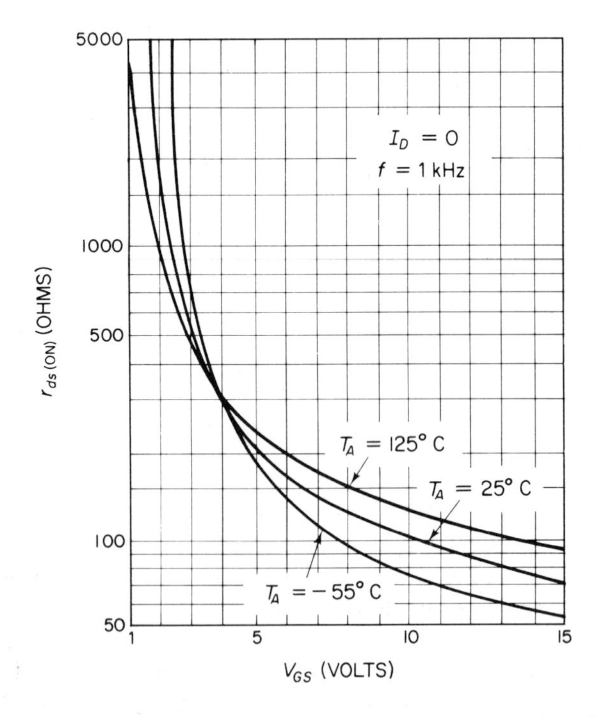

Fig. 2-85 Drain-source ON resistance (Courtesy of Motorola Inc., Semiconductor Products Division)

of 10 to 20 V to achieve an R_{DS} minimum. From Fig. 2-84, at $V_{GS} = -10$ V and temperature 25°C, R_{DS} is 300 ohms.

Compare this with Fig. 2-85 which shows the same characteristic for the MM2102 (an *N*-channel IGFET). Note that for $V_{GS} = 10$ V and temperature 25°C, R_{DS} is 100 ohms.

One reason for this 3:1 improvement in R_{DS} is that in the *P*-channel device, the carriers are holes, while in the *N*-channel FET, the carriers are electrons. The mobility of electrons is greater than that of holes, and thus is responsible for part of the improvement in R_{DS}. Since a low R_{DS} is needed in the ON condition, the *N*-channel IGFET is preferable for chopper and analog switching applications.

It is very important to know how much capacitance must be charged and discharged during transition times. Figure 2-86 shows a plot of the small-signal, common-source, short-circuit input capacitance C_{iss}, and reverse transfer capacitance C_{rss} versus voltage of the *N*-channel and *P*-channel IGFETs. There is no appreciable change in either capacitance with voltage.

As shown in Fig. 2-87, the capacitance of the JFET does vary with voltage (approximately the square root of voltage). Figure 2-87 is a valuable design aid in determining the capacitance at a particular operating point.

C_{rss} is the capacitance from gate-to-drain, and is the capacitance that causes the feedthrough of the control signal to the load. C_{iss} is the parallel combination of gate-to-drain and gate-to-source capacitance C_{gd} and G_{gs}.

C_{gs} and C_{gd} will together form a series capacitance in parallel with R_{DS}. When the FET is used as an analog switch, this capacitance will bypass R_{DS} at high input frequencies. This bypass will therefore limit the frequency at which the FET can be used as an analog switch.

For example, note that the C_{iss} capacitance is greater than 1 pF at any V_{GS}. At a frequency of 100 MHz, the reactance of a 1-pF capacitor is about 1 kΩ. If the R_{DS} is greater than 1 kΩ, the 100-MHz signals will be bypassed

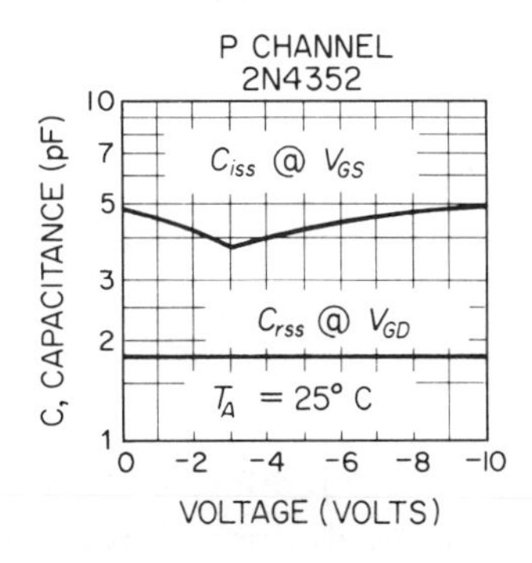

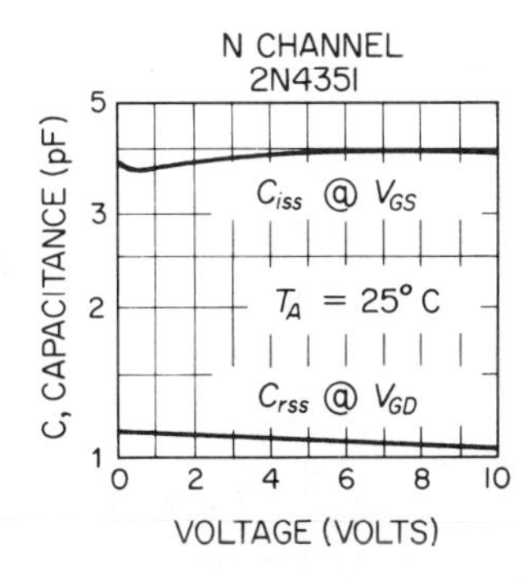

Fig. 2-86 C_{iss} and C_{rss} of IGFETs (Courtesy of Motorola Inc., Semiconductor Products Division)

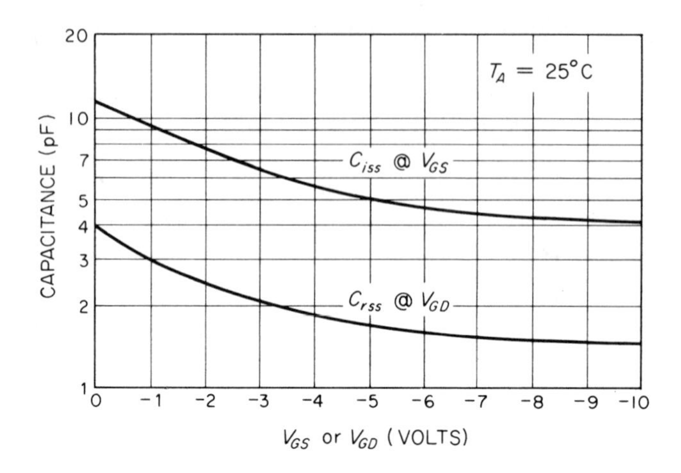

Fig. 2-87 3N126 capacitance versus voltage (Courtesy of Motorola Inc., Semiconductor Products Division)

around R_{DS}. If the capacitance is increased to 10 pF, the reactance drops to about 100 ohms. This will bypass most FETs in the OFF condtion.

2-23-2 *Basic* FET *Chopper Circuits*

There are three basic chopper configurations: the series chopper, the shunt chopper, and the series-shunt chopper.

Series chopper. The series chopper is the most commonly used. The basic circuit, equivalent circuit, and equations for the series chopper are shown in Fig. 2-88. In order to operate in the ohmic region when the FET is ON, the drain current must be limited to a low value. Ordinarily, a large value of load resistance is used to limit the current. In the case of the series chopper, a high value of load resistance also minimizes the ON voltage error due to R_{DS}. The ON voltage error is given by

$$E_{\text{ON error}} = \frac{E_s(R_s + R_{DS})}{R_s + R_{DS} + R_L}$$

When the FET is OFF, a small error due to leakage of the FET is present. This OFF error is given by

$$E_{\text{OFF error}} = I_{DG} R_L$$

where $I_{DG} = I_{D(\text{OFF})}$.

The leakage of the FET (and in particular of the IGFET) is low, and the resultant OFF voltage error will be very small. Typically, for an IGFET chopper with $R_L = 100\ \text{k}\Omega$, the OFF error isl ess than a microvolt at room temperature.

FET SERIES CHOPPER

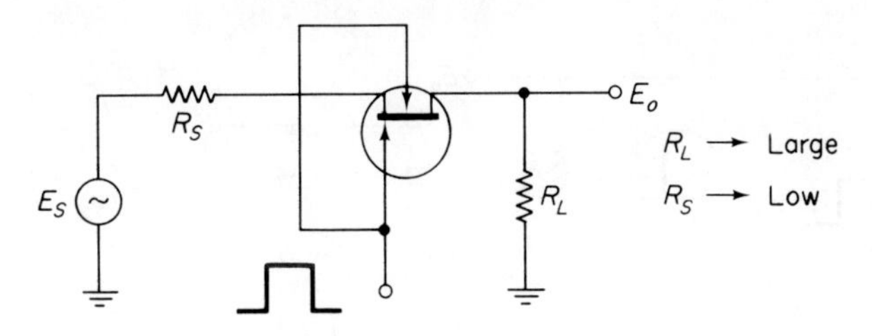

EQUIVALENT ON CIRCUIT

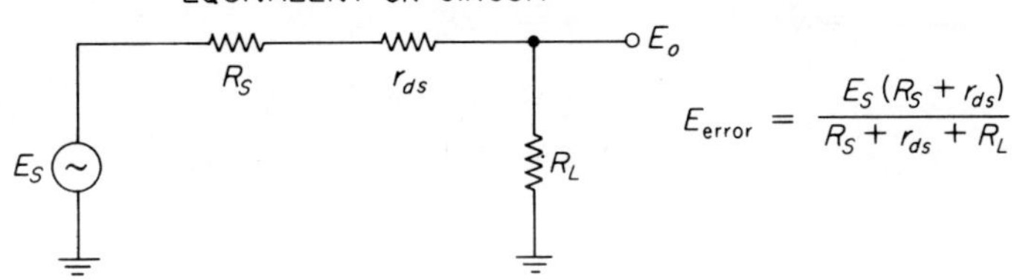

EQUIVALENT OFF CIRCUIT

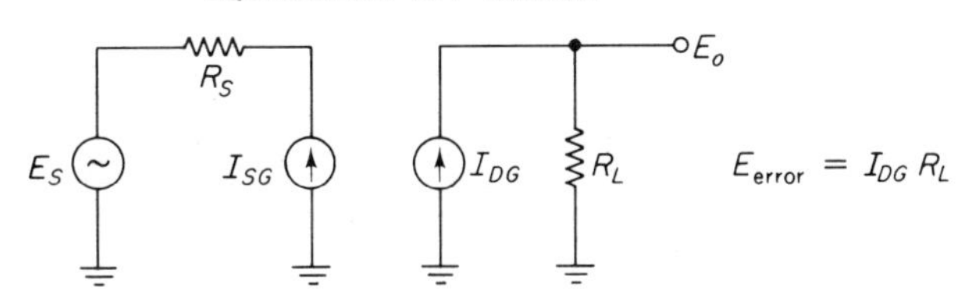

Fig. 2-88 FET series chopper equivalent circuits (Courtesy of Motorola Inc., Semiconductor Products Division)

Shunt chopper. The simple shunt chopper shown in Fig. 2-89 performs the chopping function by periodically shorting the input to ground. From the ON equivalent circuit, Fig. 2-89b, note that the shunt circuit is advantageous where a large source resistance R_s is present. The ON voltage error is given by

$$E_{\text{ON error}} = \frac{E_s R_{DS}}{R_{DS} + R_s}$$

where $R_L \gg R_{DS}$.

When the FET is OFF, the leakage current again produces an error voltage given by

$$E_{\text{OFF error}} = \frac{I_{DGO} R_s R_L}{R_L + R_s}$$

However, the OFF error is usually small compared to the drop across R_s.

Series-shunt chopper. The series-shunt chopper, Fig. 2-90, operates on the following principle: When Q_1 is ON, Q_2 is OFF, and conversely, when Q_1

FET SHUNT CHOPPER

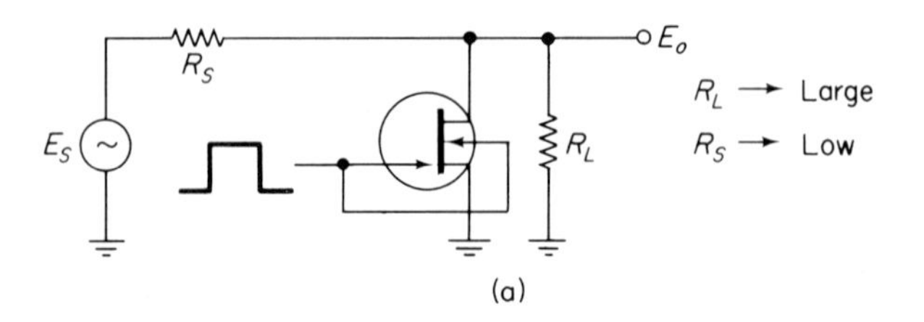

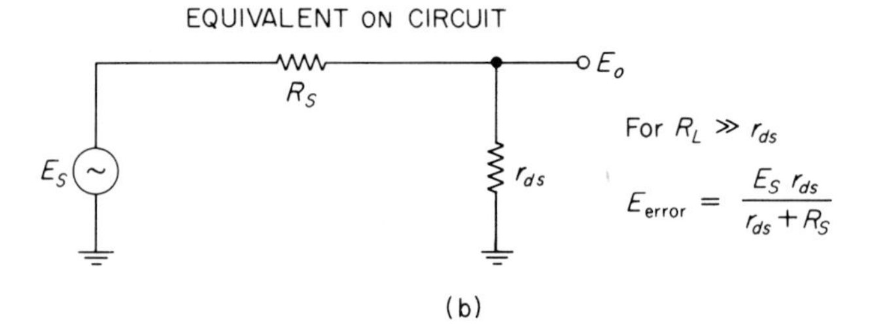

EQUIVALENT OFF CIRCUIT

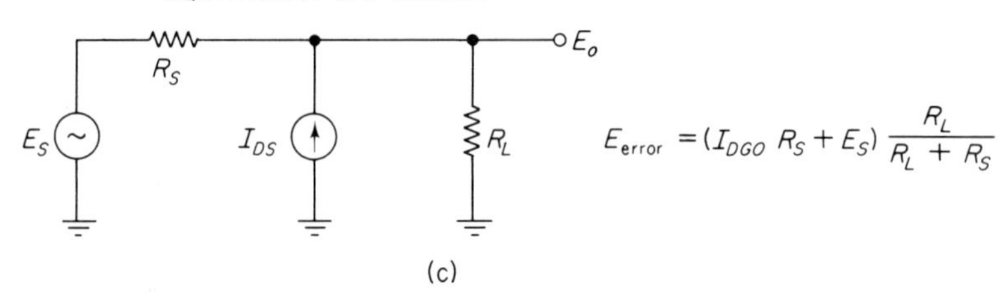

Fig. 2-89 FET shunt chopper equivalent circuits (Courtesy of Motorola Inc., Semiconductor Products Division)

is OFF, Q_2 is ON. The equivalent ON-OFF circuits are also shown in Fig. 2-90. When Q_1 is ON and Q_2 is OFF, the output is similar to that of the series chopper, except for the small error introduced by the leakage of Q_2:

$$E_{\text{ON error}} = \frac{R_L[E_s + I_{DGO}(R_s + R_{DS})]}{R_L + R_s + R_{DS}}$$

When Q_1 is OFF and Q_2 is ON, the OFF voltage error due to the leakage of Q_1 is reduced since R_{DS2} appears in parallel with R_L. This can be seen from the OFF voltage error equation:

$$E_{\text{OFF error}} = \frac{I_{DGO} R_{DS} R_L}{R_L + R_{DS}}$$

However, the leakage current of an IGFET is quite small, so the series-shunt circuit cannot be justified for the sake of minimizing the error due to leakage. The series-shunt circuit does have a definite advantage in the area of high-frequency chopping.

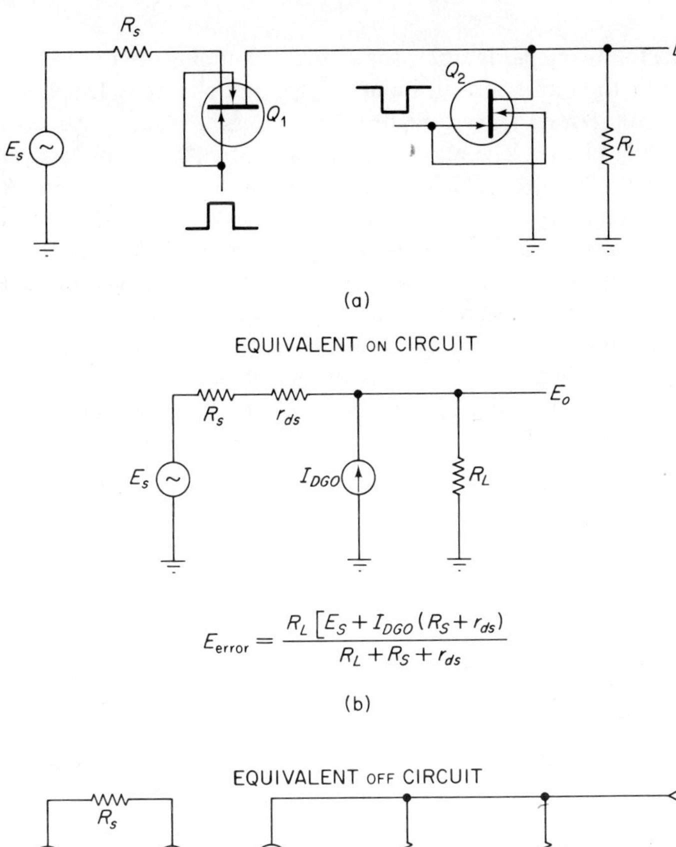

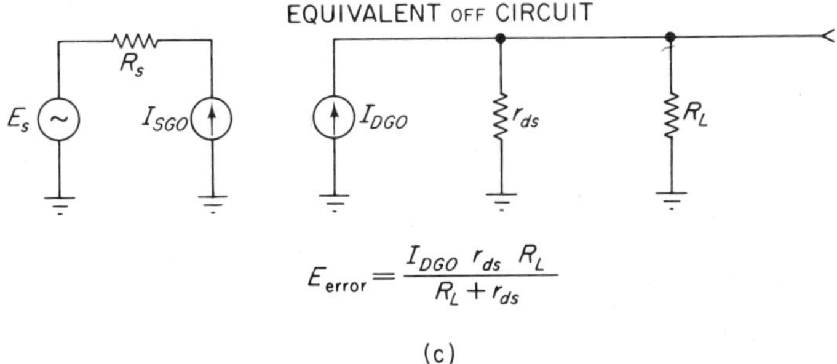

Fig. 2-90 FET series-shunt chopper equivalent circuits (Courtesy of Motorola Inc., Semiconductor Products Division)

In the simple series chopper, when the FET is OFF, C_{rss} must be discharged through the load resistor R_L. The relatively long-time constant ($C_{rss} R_L$) will limit the chopping frequency. In the series-shunt chopper, however, every time the series device turns OFF, the shunting device is turned ON, and the low resistance of Q_2 will parallel R_L. The RC time constant will, therefore, be greatly reduced, and the chopping frequency can be increased significantly.

2-23-3 *Practical* FET *Chopper Circuits*

In the following paragraphs, some actual chopping circuits are examined to determine the limitations of input voltage and chopping frequency.

Practical series chopper. Figure 2-91 shows a simple, but complete, series chopper circuit. The maximum chopping frequency for this particular circuit is about 200 kHz. This limitation is primarily due to the long RC time constant for discharging $C_{rss}C_{gd}$ through the 10 kΩ resistors. The time constant can be shortened, and the frequency increased, if the 10 kΩ resistor is reduced. However, this results in a greater voltage drop across the divider consisting of the drain-source resistance and the load resistor.

The maximum allowable input voltages are +2 V and −0.4 V. The reason for these voltage limitations is the ON-OFF requirement of the JFET. For the particular N-channel JFET shown (3N126), the ON condition is 0 V, and the OFF condition is $V_{GS} > 4$ V. The ON condition is obtained by grounding the gate. When the source (E_S, supply) starts to go positive, there is a negative potential from gate to source. The negative gate-source voltage causes R_{DS} to increase, and the FET starts to turn OFF.

When the input (source) goes more negative than −0.4 V, the P-N diode from gate to source becomes forward-biased, and the FET starts to turn ON. Hence, the negative limitation of −0.4 V.

In addition to these positive and negative input voltage limitations, there is a minimum input voltage limitation, due to feedthrough spikes in the output channel. This feedthrough is caused by C_{rss} and C_{iss}, as previously discussed. For inputs less than about 10 mV, the feedthrough spikes become an appreciable part of the output waveform (particularly at high frequencies).

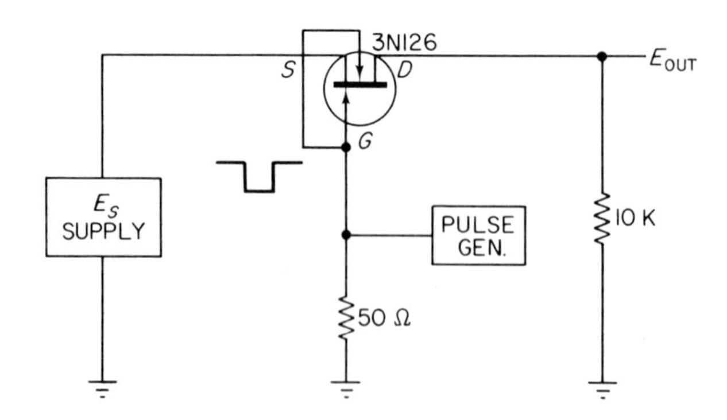

MAXIMUM CHOPPING FREQUENCY $f_{(max)} \simeq 200$ kHz
MAXIMUM INPUT VOLTAGE $E_{S(max)} \simeq +2$ V, −0.4 V

Fig. 2-91 Practical series chopper using N-channel JFET (Courtesy of Motorola Inc., Semiconductor Products Division)

There are several circuit techniques useful for minimizing these spikes. First, the control signal (pulse generator) at the gate can be a "sloppy" square wave. That is, the dv/dt of the input pulse should be kept as low a value as possible. Sharp corners on the waveform should be avoided. (A sine wave could be used.) Next, a capacitor can be connected across the output to filter the spikes. Finally, if a fixed amplitude output is acceptable, a clipper circuit can be connected across the output.

Modified series chopper for large input voltages. When a FET is OFF, large negative input voltage will tend to turn the FET ON again. The circuit of Fig. 2-91 can be modified to get around this input voltage limitation. For example, in order to overcome the limitation of maximum input voltage when the FET is turned ON, the circuit of Fig. 2-92 can be used. Operation of this circuit is as follows.

When the FET is ON, the driver transistor is OFF, and a potential of +10 V appears at diode D_1. This reverse-biases D_1. When the input (source) goes positive, the 100 kΩ resistor from source to gate makes the gate follow in potential as long as the diode does not become forward-biased. If the diode D_1 starts to conduct, a negative potential appears from gate to source, and the FET starts to turn OFF. For negative input-voltages, there is no problem as before, since the gate will follow the source until D_1 avalanches.

The circuit of Fig. 2-92 also improves the performance when the FET is OFF. Under these conditions, the driver transistor is turned ON, and when the gate of the FET is at −15 V, the FET will remain OFF for inputs up to −10 V.

The maximum input voltage of ±10 V is a function of the bias voltage. The real limitation for the input voltage to this circuit is the source-to-gate breakdown voltage. Typical breakdown voltage is 50 V for a JFET and 30 V for an IGFET. With different values of bias, the input could be increased

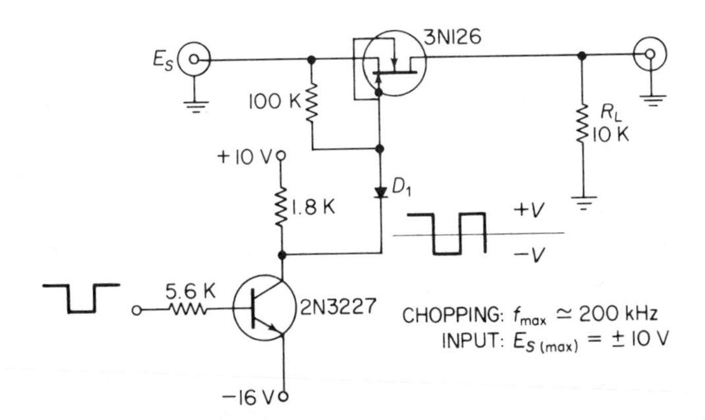

Fig. 2-92 JFET circuit for large input voltages (Courtesy of Motorola Inc., Semiconductor Products Division)

to about +22 V (when both positive and negative inputs are applied), or to about +44 V (when only positive inputs are used. Approximately −6 V is required to keep the 3N126 OFF.)

Practical series-shunt chopper for high-frequency use. Figure 2-93 is a series-shunt high-frequency chopper using complementary enhancement-mode IGFETs. Using the components shown, the circuit will operate satisfactorily at frequencies up to about 5 MHz. An *N*-channel and a *P*-channel IGFET are used as the series and shunting devices, respectively. This allows one drive circuit for both devices.

When the series FET is OFF, the R_{DS} of the shunting FET is about 200 ohms. This value parallels R_L to ground, and reduces the net output load resistance to about 200 ohms (the parallel combination of 200 ohms and 10 kΩ). Thus, the *RC* time constant is reduced to 2 per cent of its original value.

The circuit of Fig. 2-93 can also be modified to accept large values of input voltage. The procedure is described in Sec. 2-23-4.

Practical series-shunt chopper for low input voltages. A series-shunt chopper capable of low-level chopping is shown in Fig. 2-94. Two *N*-channel IGFETs with *matched* C_{rss} are used in the circuit. The gate drives for the pair are produced by a current-mode astable multivibrator. A current-mode multivibrator is one where high switching currents are used. This results in good frequency stability. The main reason for using a current-mode multivibrator here is so that the complementary outputs of the FETs are not delayed in time, with respect to each other. That is, when one output is turning OFF, the other output must be concurrently turning ON.

By matching C_{rss}, the feedthrough spikes in the output of the chopper can be nearly eliminated. Complete elimination of feedthrough spikes is difficult to obtain since the turn-on and turn-off characteristics of IGFETs are not symmetrical.

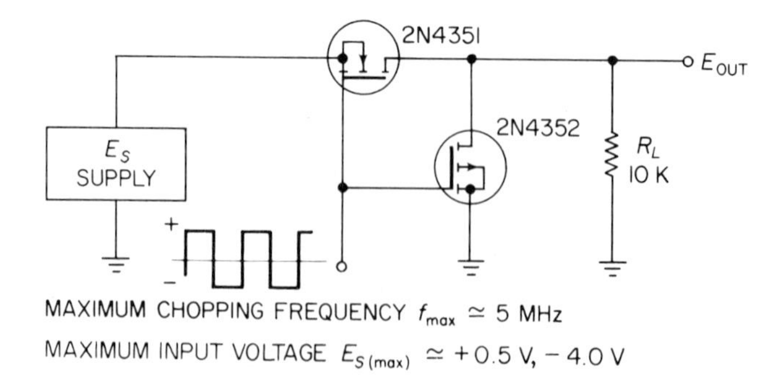

Fig. 2-93 Series-shunt chopper for high-frequency applications using complementary enhancement mode IGFETs (Courtesy of Motorola Inc., Semiconductor Products Division)

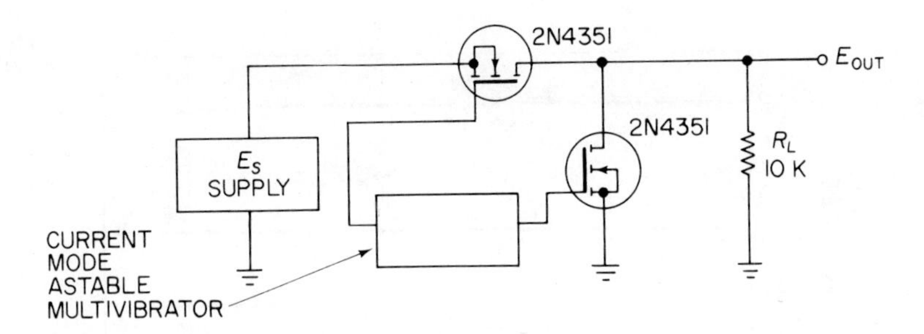

MAXIMUM CHOPPING FREQUENCY $f_{(max)} \simeq 5$ MHz

MAXIMUM INPUT VOLTAGE $E_{S(min)} \simeq \pm 10\ \mu V$

Fig. 2-94 FET series-shunt chopper for low input voltages (Courtesy of Motorola Inc., Semiconductor Products Division)

2-23-4 FET *Analog Switching Circuits*

Most of the information described thus far for FET choppers can be applied to FET analog switches. By definition, the analog switch is a device that either transmits an analog signal without distortion, or completely blocks it off.

Typically, an *N*-channel JFET is capable of passing input frequencies up to 20 MHz, without appreciable distortion or attenuation.

The frequency response curves for a 3N126 JFET used as an analog switching device are shown in Fig. 2-95. In running these curves, the JFET was operated as a switch between a sine-wave input voltage of 0.2 V, and an *RF* voltmeter monitoring the output. Readings of the output voltage were taken for two values of load resistance R_L. The output was measured both with the FET ON and OFF.

For a load resistance $R_L = 10\ k\Omega$ and the FET ON, the curve shows the output to be 2.5 dB down at 20 MHz. The $R_L = 10\ k\Omega$ curve for the OFF condition shows the output to be 22 dB at 20 MHz, or approximately 20 dB of separation between ON and OFF.

With a 50-ohm load, considerable loss is experienced due to the R_{DS} (approximately 500 ohms) of the FET. The output is 14 dB down at 20 MHz for the FET turned ON, and 40 dB down for the FET turned OFF, resulting in a separation of 26 dB.

The circuit of Fig. 2-92 can be used as an analog switch as well as a chopper. Such a switch is able to pass an input signal of ± 10 V, with a frequency up to 20 MHz, without appreciable distortion or attenuation.

Another analog switching circuit using an IGFET is shown in Fig. 2-96. (Of course, the circuit can also be used as a chopper.) The problem of han-

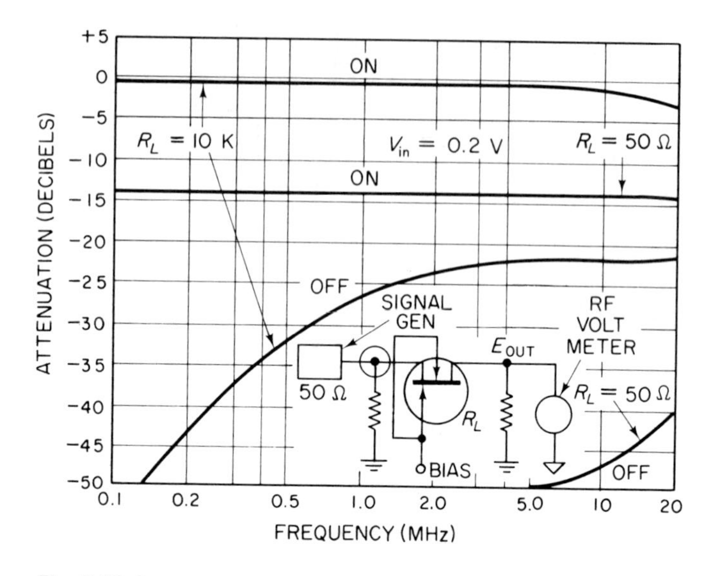

Fig. 2-95 Frequency response of 3N126 (Courtesy of Motorola Inc., Semiconductor Products Division)

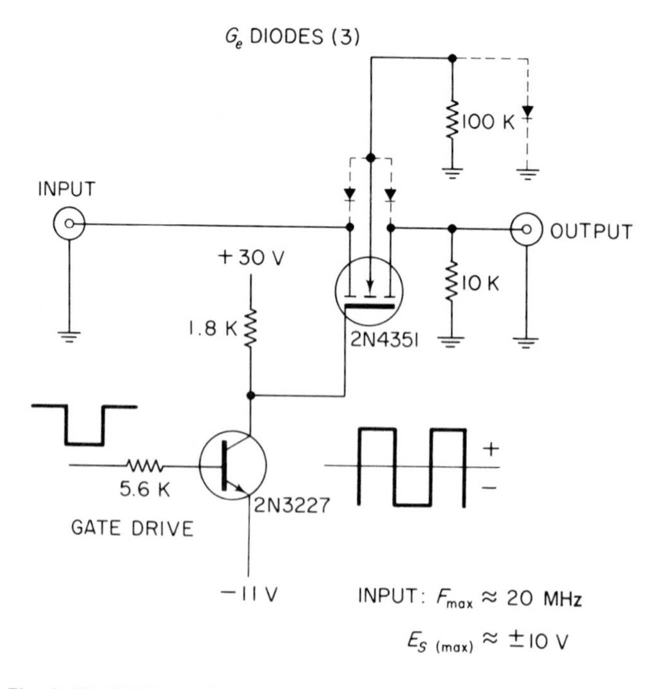

Fig. 2-96 IGFET circuit for large input voltages (Courtesy of Motorola Inc., Semiconductor Products Division)

dling large plus or minus values of input voltage is solved here in a somewhat different fashion than for the JFET circuit of Fig. 2-92. With the IGFET circuit of Fig. 2-96, there is no *P-N* junction to worry about since the gate is insulated from the rest of the FET. This eliminates the need for a diode at the gate. There are, however, *P-N* junctions from substrate-to-source, and substrate-to-drain. These junctions must not be allowed to become forward-biased. One way to accomplish this is to cut off the substrate lead, and leave it floating. However, since the substrate is connected to the can, this results in quite a bit of pick-up.

A more practical solution is shown in Fig. 2-96. Here the substrate-to-source junction is coupled with a diode. In turn, the substrate is coupled back to ground through a 100 kΩ resistor.

Commutator using FET analog switches. The analog switch can be used in a commutator circuit, as shown in Fig. 2-97. Each switch has a three-input AND gate in series with the gate drive. In order to turn a switch ON, a positive

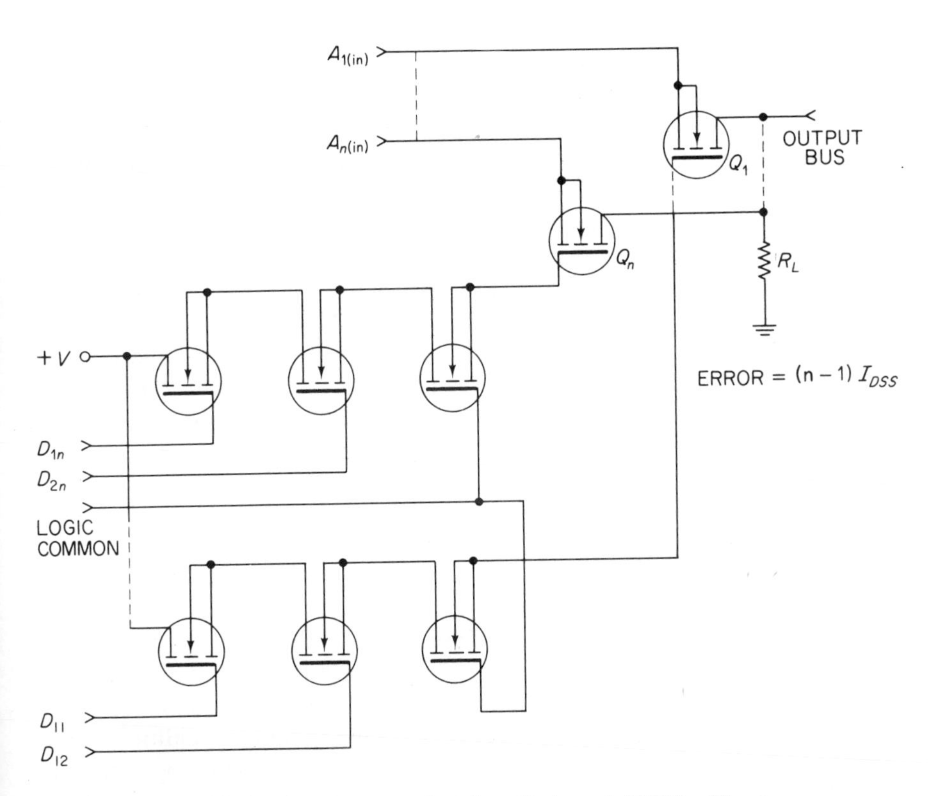

Fig. 2-97 Commutator network using *N*-channel IGFETs (Courtesy of Motorola Inc., Semiconductor Products Division)

potential is required at the gate. To accomplish this, the three inputs of the AND gate must be "true" (at a positive potential).

Assume that A_n is to be sampled. Suppose the logic common is a clock signal, that D_{1n} is an order from the control system to sample A_n, and D_{2n} is a ready signal from the device to be sampled. When all of these conditions are true at the same time, switch Q_n is turned ON and A_n is sampled. The circuit of Fig. 2-97 can be modified to accept large values of input voltage.

In this type of commutating circuitry, where only one channel is turned ON at a time, an error is introduced due to the leakages of the other FET switches. Assuming that the leakage is the same for all switches, an approximation of the error signal is given by $(n - 1) \times I_{DSS} \times R_L$.

2-24 FET Logic Circuits

Field effect transistors are being used to simplify logic circuits. Particularly useful are the MOSFETs found in integrated circuit (IC) form. Virtually any of the conventional logic forms in common use can be fabricated in the form of IC FETs. The subject of logic circuits in general, and IC logic in particular, is quite extensive and beyond the scope of this book. For detailed discussions of logic and ICs, see the author's *Handbook of Logic Circuits*, and *Manual for IC Users*, both published by Reston Publishing Company, Inc., Reston, VA.

The following is a brief summary of how MOSFETs are used in logic work, either as ICs or in discrete form.

Basic to the use of FETs in logic circuitry is the complementary inverter of Fig. 2-98. Type C (enhancement-mode only) MOSFETs are used in this circuit. The logic levels for the inverter are +V for a 1, and ground for a 0.

With a true input (+V), the *P*-channel stage has zero gate voltage, and is essentially cut off. The *P*-channel conducts very little drain current (I_{DSS}, and typically a few picoamperes for a C-type FET). The *N*-channel element is forward-biased and its drain voltage (with only a few picoamperes of I_{DSS} allowed to flow) is near ground or false (0). The load capacitance C_L represents the output load, plus any stray circuit capacitance.

With a false input (ground), the *N*-channel element is cut off and permits only I_{DSS} to flow. The *P*-channel element is forward-biased and its V_{DS} is low. Thus, the drain terminal of the *P*-channel is near +V, and C_L is charged to +V. The power dissipation is extremely low, since both stable states, true and false (or 1 and 0), are conducting only leakage current. Power is dissipated only during switching, an ideal situation for logic circuits. In addition to the lower power dissipation, another advantage of IGFETs for logic circuits is that no coupling elements are required (the gate acts as a coupling capacitor).

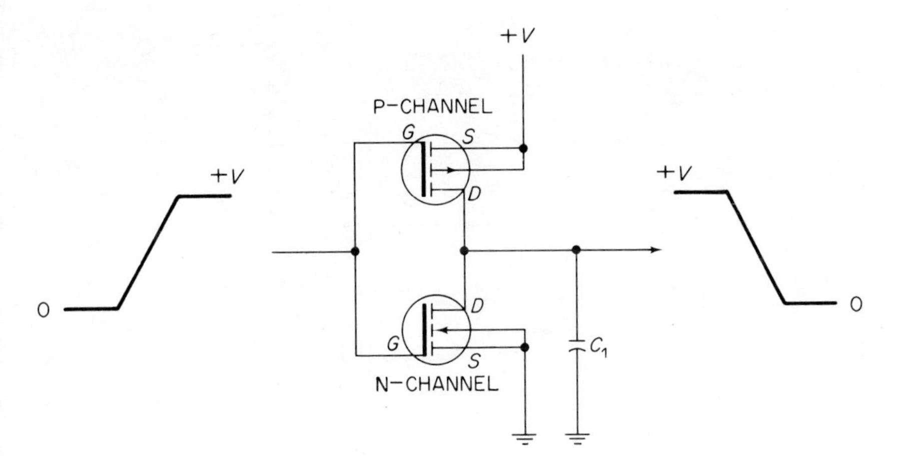

Fig. 2-98 Basic FET complementary inverter logic circuit (Courtesy of Motorola Inc., Semiconductor Products Division)

With the need for a coupling capacitor function, it is relatively simple to fabricate MOSFET logic elements in IC form (fabrication of capacitors is usually the stumbling block for most IC design). OR, NOR, AND and NAND gates with either positive or negative logic can be implemented with MOSFETs. Thus, almost any logic circuit combination can be produced. FET logic can also be used over a wide range of power-supply voltages.

An *RS* flip-flop, a two-input NOR gate, and a two-input NAND gate are shown in Figs. 2-99, 2-100, and 2-101, respectively.

In the NOR gate circuit of Fig. 2-100, the gates of Q_1 and Q_3 are tied

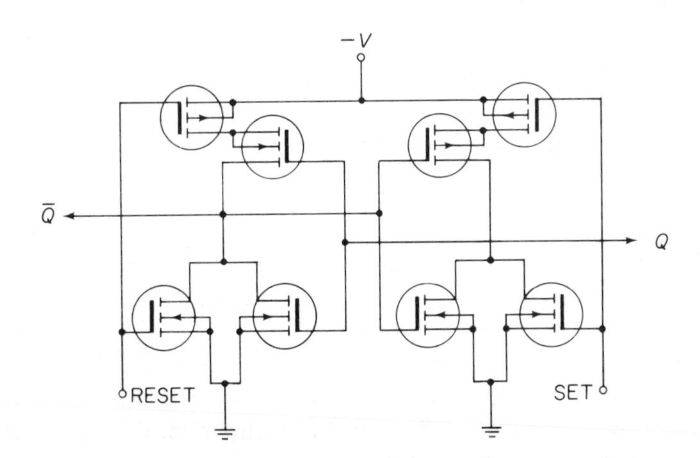

Fig. 2-99 Complementary FET R_S flip-flop (Courtesy of Motorola Inc., Semiconductor Products Division)

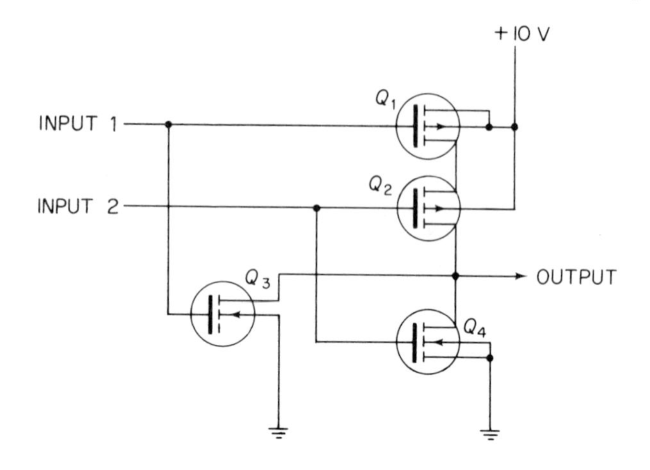

0 = 0 V		1 = + 10 V
INPUT 1	INPUT 2	OUTPUT
0	0	1
0	1	0
1	0	0
1	1	0

Fig. 2-100 Two-input NOR gate using MOSFETs (Courtesy of Motorola Inc., Semiconductor Products Division)

together to form input 1. Elements Q_1 and Q_3 act together as an inverter circuit in that they form a push-pull combination. Element Q_2 acts as a series resistance, which is either extremely high or low, in the inverter formed by Q_1 and Q_3. Likewise, Q_1 acts as a series resistance in the inverter formed by Q_3 and Q_4.

Input 2 is the control input for the second inverter. The output of the circuit is at +10 V only when elements Q_1 and Q_2 are ON. This occurs only if both inputs 1 and 2 are at ground. Thus, the output is a logic 1 only when both inputs are at a logic 0.

Since a NOR gate using positive logic becomes a NAND gate when using negative logic, the NOR gate can be converted to a NAND gate by interchanging *P*- and *N*-channel elements, and flipping the circuit upside down. This is done with the two-input NAND gate of Fig. 2-101.

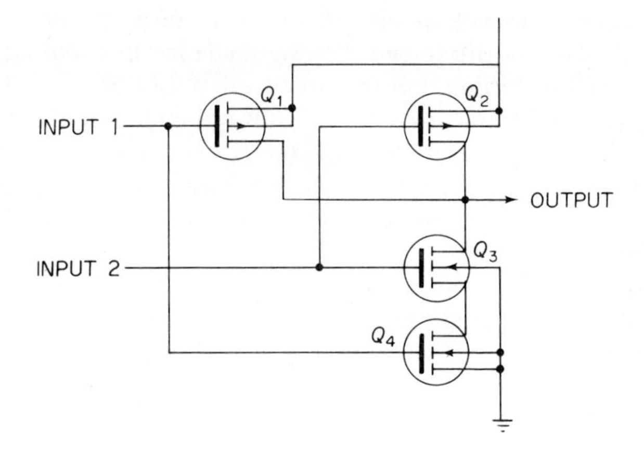

0 = 0 V		1 = + 10 V
INPUT 1	INPUT 2	OUTPUT
0	0	1
0	1	1
1	0	1
1	1	0

Fig. 2-101 Two-input NAND gate using MOSFETs (Courtesy of Motorola Inc., Semiconductor Products Division)

2-25 FET Oscillators

The characteristics of FETs are very similar to those of pentode vacuum tubes. Therefore, practically any of the classic vacuum-tube oscillators can be designed with a FET. For this reason, we will not discuss the details for all oscillator circuits. Instead, we shall discuss three basic FET oscillators, two for RF and one for audio.

The main concern in any oscillator design is that the FET will oscillate at the desired frequency, and will produce the desired voltage or power. In comparison to bipolar transistors, FETs have a higher operating frequency. However, FETs have the disadvantage of lower power output. At best, with present-day FETs, the power output is in the order of a few milliwatts.

Another problem with oscillators is the class of operation. If an oscillator is biased class A (with some I_D flowing at all times), the output waveform will be free from distortion, but the circuit will not be efficient. That is, the power output will be low in relation to power input. For the purposes of calculation, input power for a FET oscillator can be considered as the product of I_D and drain voltage. Class A oscillators are usually not used for RF, and are generally limited to those applications where a good waveform is the prime consideration.

A class C oscillator (where I_D is cut off by feedback greater than V_p) is far more efficient. This cuts both power and heat requirements. At radio frequencies, the waveform is usually not critical, so class C is in common use for RF circuits.

In the following system of design, the class of operation is set by the amount of feedback, rather than by bias point. That is, the FET is biased for an optimum operating point, and then feedback is adjusted for the desired class of operation.

2-25-1 Crystal-Controlled Oscillator

Figure 2-102 is the working schematic of a crystal-controlled oscillator. This circuit is one of the many variations of the Colpitts oscillator. However, the output frequency is fixed, and is controlled by the crystal. The circuit can be tuned over a narrow range by L_1 (which is slug-tuned). For maximum efficiency, the resonant circuit (C_1, C_2, L_1, and the FET output capacitance)

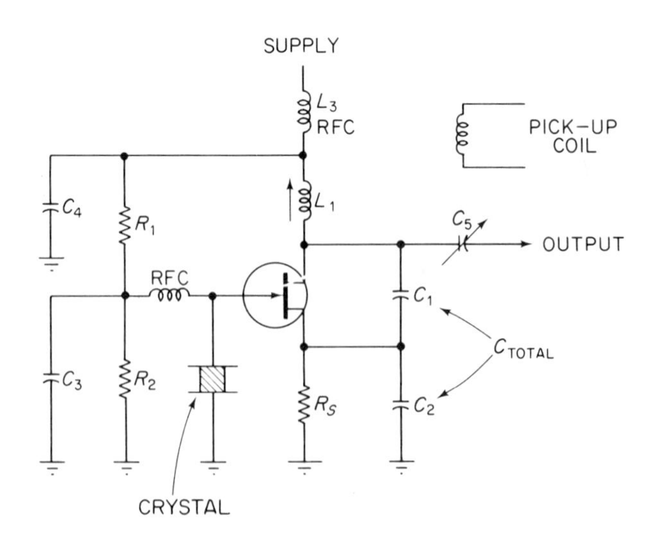

Fig. 2-102 FET crystal-controlled oscillator

should be at the same frequency as the crystal. If reduced efficiency is acceptable, the resonant circuit can be at a higher frequency (multiple) of the crystal frequency. However, the resonant circuit should not be operated at a frequency higher than the *fourth harmonic* of the crystal frequency.

Bias circuit. The bias circuit components, R_1, R_2, and R_S, are selected by using the procedures described in Secs. 2-5 and 2-8. That is, a given I_D is selected, and the bias components are chosen to produce that I_D under no-signal conditions. The bias circuit is calculated and tested on the basis of normal operating point, even though the circuit will never be at the operating point. A feedback signal is always present, and the FET is always in a state of transition. Unless there is some other special requirement, the normal operating point should be the 0TC point. This will provide the greatest stability. However, a FET oscillator can be operated at other points (different values of no-signal I_D) to meet some particular power requirement. For example, it may be necessary to operate the FET at some I_D higher than the 0TC point to obtain a higher power output.

Feedback signal. The signal output appears at the FET drain terminal. With the proper bias-feedback relationship, the output signal is about 80 per cent of the supply voltage. The amount of feedback is determined by the ratio of C_1 and C_2. For example, if C_1 and C_2 are the same value, the feedback signal is one-half of the output signal. If C_2 is made about three times the value of C_1, the feedback signal is about 0.25 of the total output signal voltage.

It may be necessary to change the value of C_1 in relation to C_2, in order to achieve a good bias-feedback relationship. For example, if C_2 is decreased in value, the feedback increases and the oscillator operates nearer the class C region. An increase in C_2, with C_1 fixed, decreases the feedback and makes the oscillator operate as class A. Keep in mind that any changes in C_2 (or C_1) will also affect frequency. Therefore, if the C_2/C_1 values are changed, it will probably be necessary to change the value of L_1.

As a first trial value, the amount of feedback should be equal to, or greater than, maximum V_P (or $V_{GS(OFF)}$). Under normal conditions, such a level of feedback should be sufficient to overcome the fixed bias (set by $R_1 - R_2$) and the variable bias set by R_S.

Frequency. Frequency of the circuit is determined by the resonant frequency of L_1, C_1, and C_2, and by the crystal frequency. Note that C_1 and C_2 are in series, so that the total capacitance (C_{total}) must be found by the conventional series equation. Also note that the output capacitance of the FET (C_{ds}) must be added to the value of C_1. At low frequencies, the output capacitance can be ignored since the value is usually quite low in relation to a typical value for C_1. At higher frequencies, the value of C_1 is lower, so the capacitance becomes of greater importance. For example, if the output capacitance is 50 pF at the frequency of interest, and the value of C_1 is 1000 pF or larger, the effect of the output capacitance will be small. (FET output capacitance can be considered as being in parallel with C_1.) If the value of C_1 is lowered to 50 pF, the parallel output capacitance will double the value. Thus, the output capacitance must be included in the resonant-frequency calculations.

Capacitor C_1 can be made variable. However, it is generally easier to make L_1 variable, since the tuning range is quite small.

Typically, the value of C_2 is about three times the value of C_1 (or the combined values of C_1 and the FET output capacitance, where applicable). Thus, the signal voltage (fed back to the FET source terminal) is about 0.25 of the total output signal voltage (or about 0.2 of the supply voltage, when the proper bias-feedback relationship is established).

Resonant circuit. Any number of L and C combinations could be used to produce the desired frequency. That is, the coil could be made very large or very small, with corresponding capacitor values. Often, practical limitations are placed on the resonant circuit (such as available variable inductance values. In the absence of such limitations, and as a starting point for resonant-circuit values, the inductive reactance of L_1 should be between 80 and 100 ohms at the operating frequency.

Output power and circuit. With the correct bias-feedback relationship, the output power of the oscillator will be about 0.3 of the input power (drain voltage or supply voltage times I_D). Output to the following stage can be taken from L_1 by means of a pick-up coil (for low-impedance loads) or coupling capacitor (for high-impedance loads). Generally, the most convenient output scheme is to use a coupling capacitor (C_5), and make the capacitor variable. This makes it possible to couple the oscillator to a variable load (a load that changes impedance with changes in frequency).

Crystal. The crystal must, of course, be resonant at the desired operating frequency (or a sub-multiple thereof when the circuit is used as a multiplier. Note that efficiency (power output in relation to power input) of the oscillator is reduced when the oscillator is also used as a multiplier. The crystal must be capable of withstanding the combined d-c and signal voltages at the FET gate. As a rule, the crystal should be capable of withstanding the full supply voltage, even though the crystal will never be operated at this level.

Bypass and coupling capacitors. The values of bypass capacitors C_3 and C_4 should be such that the reactance is 5 ohms or less at the crystal operating frequency. A higher reactance could be tolerated (200 ohms). However, due to the low crystal output, the lower reactance is preferred.

The value of C_5 should be approximately equal to the combined parallel output capacitance of the FET and C_1. Make this the mid-range value of C_5, if C_5 is variable.

Radio-frequency chokes. The values of radio-frequency chokes (RFC) L_2 and L_3 should be such that the reactance is between 1000 and 3000 ohms at the operating frequency. The minimum current capacity of the chokes should be greater (by at least 10 per cent) than the maximum anticipated direct current. Note that a high reactance is desired at the operating frequency. However, at high frequencies, this could result in very large chokes that produce a large voltage drop (or are too large physically).

CRYSTAL OSCILLATOR DESIGN EXAMPLE

Assume that the circuit of Fig. 2-102 is to provide an output at 50 MHz. The circuit is to be tuned by L_1. A 30-V supply is available. The crystal will not be damaged by 30 V, and will operate at 50 MHz with the desired accuracy. The FET to be used is the 2N5268. The bias network is to be identical to that described in Sec. 2-8, except that R_L is omitted. Thus, the FET drain is operated at 30 V (ignoring the small drop across L_1 and L_3). The values of R_1, R_2, and R_S are identical to those described in Sec. 2-8. Assume that the FET has an output capacitance (C_{ds}) of 50 pF at the operating frequency of 50 MHz.

With the drain at 30 V and an I_D of 1 mA at the "operating point," the power input is 30 mW. Assuming a typical efficiency of 0.3, the output power is about 9 mW.

With a 30-V supply, the output signal should be about 24 V ($30 \times 0.8 = 24$). Of course, this is dependent upon the bias-feedback relationship.

Assume that, the maximum V_P (or $V_{GS(\text{OFF})}$) is 6 V. Thus, feedback should be 6 V or greater. When C_2 is made three times the value of C_1, the feedback signal will be 6 V ($24\text{ V} \times 0.25 = 6\text{ V}$). Considering the amount of fixed and variable bias supplied by the bias network, a feedback of 6 V may be large. However, the 6-V value should serve as a good starting point.

As discussed, for realistic L and C values in the resonant circuit, the inductive reactance of L_1 should be between 80 and 100 ohms at the operating frequency. Assume a value of 80 ohms as a first trial.

With an inductive reactance of 80 and an operating frequency of 50 MHz, the inductance of L_1 should be $80 \div 6.28 \times (50 \times 10^6) = 0.2\ \mu\text{H}$. For convenience, the inductance should be tunable from about 0.15 to 0.3 μH.

With a value of 0.2 μH for L_1 and an operating frequency of 50 MHz, the total capacitance of C_1 (plus the parallel output capacitance of the FET) and C_2 should be $2.54 \times 10^4 \div (50{,}000)^2 \times 0.2$, or 50 pF.

With a 24-V output and a 6-V feedback, the value of C_1 (plus output capacitance) is $50 \times (6 + 24) \div 24$, or 63 pF. Since the output capacitance of the FET is 50 pF, the value of C_1 should be $63 - 50$, or 13 pF.

The value of C_2 is 63 × 3, or 189 pF (rounded to 190 pF).

Keep in mind that an incorrect bias-feedback relation will result in distortion of the waveform, or low power, or both. The final test of correct operating point is a good waveform at the operating frequency, together with the desired output power.

The values of C_3 and C_4 should be $1 \div 6.28 \times (50 \times 10^6) \times 5$, or 630 pF minimum. A slightly larger value (say, 1000 pF) will assure a reactance of less than 5 at the operating frequency.

The values of L_2 and L_3 should be $2000 \div 6.28 \times (50 \times 10^6)$, or 6.3 μH nominal. Any value between about 3 and 9 μH should be satisfactory. The best test for the correct value of an RF choke is to check for RF at the power supply side of the line, with the oscillator operating. There should be no RF, or the RF should be a fraction of 1 V (usually less than a few microvolts for a typical FET oscillator). If RF is removed from the powers supply line, the choke reactance is sufficiently high. Next, check for d-c voltage drop across the choke. The drop should be a fraction of 1 V (also in the microvolt range).

2-25-2 Variable-Frequency Oscillator

Figure 2-103 is the working schematic of a variable-frequency oscillator. This circuit is also one of the many variations of the Colpitts oscillator. The circuit is chosen for maximum stability at frequencies up to about 0.5 MHz. Oscillation is sustained by source feedback (from the junction of C_1 and C_2), rather than by gate feedback.

Design considerations. All of the design considerations for the variable-frequency oscillator are the same as for the crystal-controlled oscillator (Sec. 2-24-1), with the following exceptions.

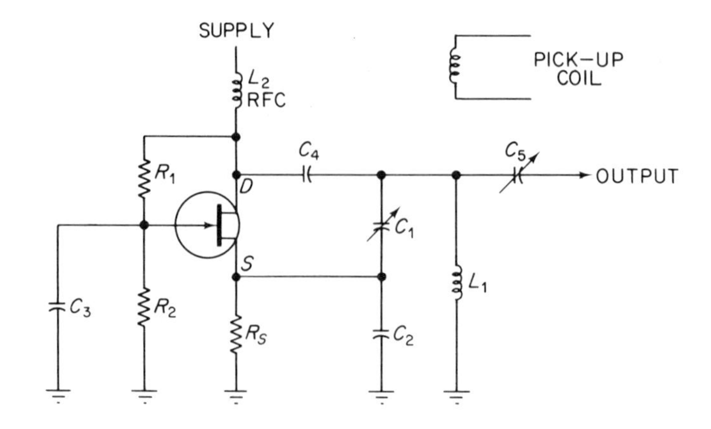

Fig. 2-103 FET variable-frequency oscillator

Generally, C_1 is made variable to tune across a given frequency range. However, L_1 can be made variable if required.

The values of coupling and bypass capacitors C_3, C_4, and C_5 (if used) should be such that the reactance is 200 ohms at the *lowest* operating frequency (when variable capacitor C_1 is at full value). Note that capacitor C_4 and the output capacitance of the FET may add to the C_1 capacitance. This tends to lower the resonant frequency of the L_1, C_1, C_2 circuit slightly from the calculated value. However, since C_1 is variable, there should be no problem in tuning to a desired frequency.

Variable-Frequency Oscillator Design Example

Assume that the circuit of Fig. 2-103 is to tune across a frequency range from about 10 kHz to 60 kHz. A 30-V power supply is available. The FET to be used is the 2N5268. The bias network is to be identical to that described in Sec. 2-8, except that R_L is omitted. Thus, the FET drain is operated at about 30 V (ignoring the small drop across L_2). The values of R_1, R_2 and R_s are identical to those described in Sec. 2-8. Assume that the FET has a negligible output capacitance (in relation to C_1) at the operating frequency.

With the drain at 30 V and I_D of 1 mA at the "operating point," the power input is 30 mW. Assuming a typical efficiency of 0.3, the output power is about 9 mW.

With a 30-V supply, the output signal should be about 24 V ($30 \times 0.8 = 24$). Of course, this is dependent upon the bias-feedback relationship.

Assume that the maximum V_P (or $V_{GS(OFF)}$) is 6 V. Thus, feedback should be 6 V or greater. When C_2 is made three times the value of C_1, the feedback signal will be 6 V ($24\text{ V} \times 0.25 = 6\text{ V}$). Considering the amount of fixed and variable bias supplied by the bias network, a feedback of 6 V may be large. However, the 6-V value should serve as a good starting point.

For realistic L and C values in the resonant circuit, the inductive reactance of L_1 should be between 80 and 100 ohms at the operating frequency. Assume a value of 100 ohms as a first trial.

With an inductive reactance of 100, and a low-frequency limit of 10 kHz, the inductance of L_1 should be $100 \div 6.28 \times (10 \times 10^3)$, or approximately 2 mH.

With a value of 2 mH for L_1 and a low-frequency limit of 10 kHz, the total capacitance of C_1 and C_2 (with the variable C_1 at its high limit) should be $2.54 \times 10^4 \div (10)^2 \times 2000$, or about 0.12 μF.

With a 24-V output and a 6-V feedback, the value of C_1 is $0.12 \times (6 + 24) \div 24$, or 0.15 μF. The value of C_2 is $0.15 \times 3 = 0.45$, which is rounded to 0.5 μF.

Keep in mind that an incorrect bias-feedback relation will result in distortion of the waveform, or low power, or both. The final test of correct operating point is a good waveform at the operating frequency, together with the desired output power.

The values of C_3, C_4, and C_5 (if used) should be $1 \div 6.28 \times (10 \times 10^3)$ 200, or about 0.08 minimum. A slightly larger value (say, 0.1 μF) will assure a reactance of less than 200 ohms at the lowest frequency.

2-25-3 RC *Phase-Shift Oscillator*

Figure 2-104 is the working schematic of an RC (resistance-capacitance) phase-shift oscillator. Such oscillators are used at audio frequencies instead of the LC (inductance-capacitance) oscillators described in Secs. 2-25-1 and 2-25-2. RC oscillators avoid the use of inductances, which are not practical in the audio-frequency range. RC oscillators are usually operated in class A,

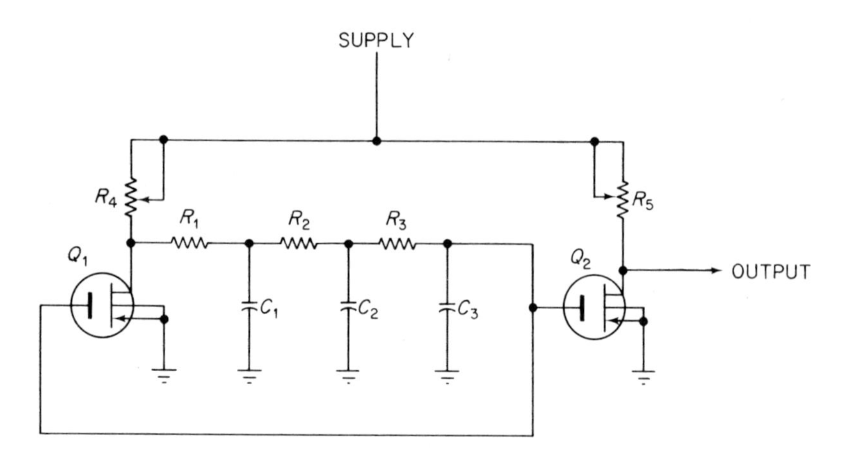

Fig. 2-104 FET *RC* phase-shift oscillator

thus producing good waveforms. The feedback principle is used in RC oscillators. In the circuit of Fig. 2-104, the output (drain) of Q_1 is fed through three RC networks back to the gate of Q_1. Each network shifts the phase about 60°, resulting in an approximate 180° shift between drain and gate. Since the drain is normally shifted 180° from the gate, the RC shift of 180° brings the feedback to 360°, or back in-phase to produce oscillation. Q_2 is used as an output amplifier.

Note that IGFETs are used for both Q_1 and Q_2. This eliminates the need for coupling capacitors between stages. Also, both IGFETs are operated without bias.

Bias network. Since both IGFETs are operated at zero gate voltage, the

Q-point drain voltage is set by I_{DSS} and the values of R_4 and R_5. As shown in Fig. 2-104, both R_4 and R_5 are made variable. Resistor R_4 is adjusted to produce oscillations with good waveforms. Resistor R_5 is adjusted to produce the desired output swing. Typically, both Q_1 and Q_2 should be operated at one-half the supply voltage. For example, if I_{DSS} is 1 mA, and the supply is 30 V, both R_4 and R_5 should be 15 kΩ, thus dropping both drains to about 15 V.

Output frequency. The oscillator frequency is determined by the RC time constant. To simplify design, the same values could be used in all three RC networks. However, such an arrangement will create a problem of power loss. Each of the RC networks functions as a low-pass filter. If the same values are used in all three sections, the signal loss will be about 15 dB through the networks. This loss, combined with the normal loss, could be sufficient to prevent oscillation if the voltage gain of Q_1 is low. The loop gain of any oscillator must be at least 1 (or slightly more for practical design). If the gain of Q_1 is 10 and the loss is anything greater than about 8 to 8.5, the circuit will not oscillate.

The RC network loss problem can be minimized by making the impedance of the succeeding network *greater* than that of the prior network. That is, there should be an impedance step-up as the signal passes from the drain of Q_1 to the gate of Q_2. For example, R_2 should be three times that of R_1; R_3 should be three times that of R_2. Thus, each RC network places very little load on the previous section and keeps loss at a minimum.

There should also be an impedance step-up between the output of Q_1 (set by R_4) and the first RC network. As a first trial, R_1 should be about 4 times the value of R_4.

The values of C_1, C_2, and C_3 must be selected to produce the desired operating frequency. The output frequency is about equal to 1/3RC. A more exact frequency calculation can not be made in practical design, since FET capacitance and resistance values must be added to the RC networks. However, the 1/3RC relationship is satisfactory for trial values.

The circuit of Fig. 2-104 is best suited to fixed-frequency use. It is difficult to find a three-section variable capacitor that will track properly.

RC Phase-Shift Oscillator Design Example

Assume that the circuit of Fig. 2-104 is to provide an output at 3.7 kHz. The power supply is 30 V. The output signal is to be the maximum possible without distortion. The IGFETs have a zero gate voltage I_D of 1 mA.

For maximum output voltage swing, the drains of both Q_1 and Q_2 should be at one-half the supply or 15 V.

With 1 mA of I_D flowing, the drops across R_4 and R_5 should be 15 V. Thus, R_4 and R_5 should be 15 kΩ (15 V/0.001 = 15 kΩ).

In practice, after the values have been selected and the components assembled, the gate of Q_2 is monitored on an oscilloscope. R_4 is adjusted for maximum signal swing without distortion. R_5 is then adjusted for maximum output swing without distortion, at the drain of Q_2.

With R_4 at 15 kΩ, R_1 should be 45 kΩ. With R_1 at 45 kΩ, and a 3.7 kHz operating frequency, C_1 should be 0.002 μF (1/3 × 45 kΩ × 3.7 kΩ = 0.002).

With R_1 at 45 kΩ, R_2 should be 135 kΩ (to achieve the impedance step-up). With R_2 at 135 kΩ, and a 3.7-kHz operating frequency, C_2 should be ≈ 0.0007 μF (1/3 × 135 kΩ × 3.7 kΩ ≈ 0.0007).

With R_2 at 135 kΩ, R_3 should be 405 kΩ. With R_3 at 405 kΩ, and a 3.7-kHz operating frequency, C_3 should be ≈ 0.00022 μF, or 2200 pF (1/3 × 405 kΩ × 3.7 kΩ ≈ 2200 pF).

2-26 Basic FET RF Amplifiers

FET RF amplifiers can be designed by using two-port networks similar to those of bipolar transistors. Basically, the method consists of characterizing the FET as a linear active two-port network (LAN) with admittances (y parameters), and using these parameters to solve exact design equations for stability, gain, and input/output admittances.

It is difficult, at best, to provide a simple, step-by-step procedure for designing FET RF amplifiers to meet all possible circuit conditions. In practice, the procedure often results in considerable trial and error. There are several reasons for this problem of FET RF amplifier design.

First, not all of the FET characteristics are always available in data-sheet form. For example, input and output admittance may be given for some low frequency, but not at the frequency of interest (frequency at which the amplifier is to be operated).

Often, manufacturers do not agree on terminology. A classic example of this is in y parameters, where one manufacturer uses letter subscripts (y_{fs}) and another manufacturer uses number subscripts (Y_{21}). Of course, this can be solved by conversions, as described later in this section.

In some cases, manufacturers will give the required information on data sheets, but not in the required form. For example, instead of giving input admittance in mhos, the input capacitance is given in farads. The input admittance is found when the input capacitance is multiplied by $6.28f$ (where f is the frequency of interest). This is based on the assumption that the input admittance is primarily capacitive, and thus dependent upon frequency. The assumption is not always true for the frequency of interest. Thus, it may be necessary to make actual tests of the FET, using complex admittance measuring equipment.

The input and output tuning circuits of an RF amplifier must perform two functions. Obviously, the circuits (capacitors and coils) must tune the amplifier to the desired frequency. The circuits must also match the input and output impedances of the FET to the impedances of the source and load. Otherwise, there will be considerable loss of signal. (Note that although impedances are involved, admittances are used instead, since admittances greatly simplify the measurement and calculations.)

Finally, as is the case with a vacuum-tube amplifier or bipolar transistor common-emitter amplifier, there is some feedback between output and input of a common-source FET RF amplifier. If the admittance factors are just right, the feedback will be of sufficient amplitude and of proper phase to cause oscillation of the amplifier. The amplifier is considered as *unstable* when this occurs. The condition is undesirable, and can be corrected by feedback (neutralization) or by changes in the input/output tuning networks. Although the neutralization and tuning circuits are relatively simple, the equations for determining stability (or instability) and impedance matching are long and complex. Generally, such equations are best solved by computer-aided design methods.

In an effort to cut through this maze of information and complex equations, we shall discuss all of the steps involved in FET RF design. Armed with this information, the reader should be able to interpret data-sheet or test information, and use the information to design tuning networks that will provide stable RF amplification at the frequencies of interest. With each step we shall discuss the various alternative procedures and types of information available. Specific design examples are given at the end of this section. These examples summarize the information contained in the various steps. On the assumption that all readers may not be familiar with two-port networks, we shall start with a summary of the y-parameter system.

2-26-1 y-Parameters

Impedance (Z) is a combination of resistance (the real part) and reactance (the imaginary part). Admittance (y) is the reciprocal of impedance, and is composed of conductance (the real part) and susceptance (the imaginary part). A y-parameter is an expression for admittance in the form:

$$y_{is} = g_{is} + jb_{is}$$

where g_{is} is the real (conductive) part of common-source input admittance, b_{is} is the imaginary (susceptive) part of common-source input admittance.

Since a FET can be treated as a linear active two-port network in small-signal applications, all of the standard y-parameter stability criteria and design procedures are directly applicable. The y-parameters are extremely

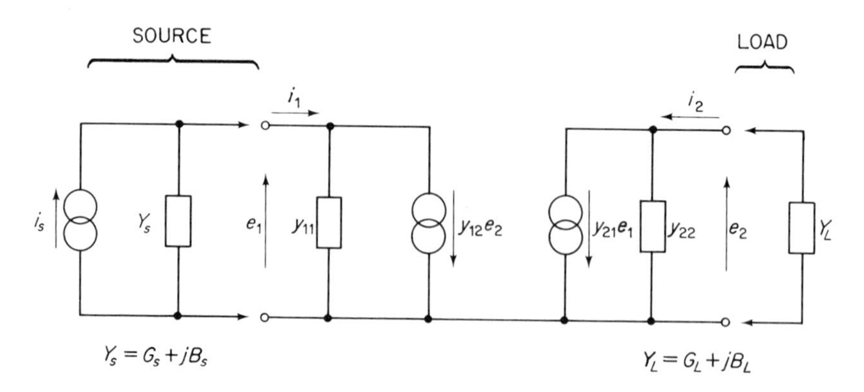

Fig. 2-105 FET y-equivalent circuit with source and load

useful in comparing different device types, in choosing a particular configuration (common-source, common-gate, neutralized, unneutralized, cascade, etc.), and in the final design of the RF amplifier.

The y-equivalent circuit is shown in Fig. 2-105. The following is a summary of the four y-parameters of primary interest.

Note that RF designers have traditionally used the nomenclature y_{11}, y_{12}, y_{21}, and y_{22} for all active devices—bipolar transistors, integrated circuits, and other devices. Some manufacturers still do. Other manufacturers use descriptive letter subscripts y_{is}, y_{rs}, y_{fs} and y_{os} for the same parameters. Both systems are given in the following summary. (Note that the letter s refers to common-source operation.)

Input admittance, with Y_L = infinity (short circuit), is expressed as

$$y_{is} = y_{11} = g_{11} + jb_{11} = y_i = \left.\frac{\Delta i_1}{\Delta e_1}\right|_{e_2=0}$$

Note that the data sheet of Fig. 2-13 does not show y_{is} or y_{11} at any frequency. However, input capacitance c_{iss} is given as 6 pF maximum. If we assume that the input admittance is entirely (or mostly) capacitive (jb_{11}), then the input impedance can be found when c_{iss} is multiplied by $6.28F$ (F = frequency in Hz). For example, if the frequency is 100 MHz and the c_{iss} is 6 pF, the input impedance is $6.28 \times (100 \times 10^6) \times (6 \times 10^{-12}) \approx 3.8$ mmhos. The assumption is accurate only if the real part of y_{is} (or g_{is}) is negligible.

Figure 2-106 shows input admittance curves for a typical FET. Note that the imaginary part (jb_{is}) is the more significant factor across the entire frequency range.

Forward transadmittance, with Y_L = infinity (short circuit), is expressed as

$$y_{fs} = y_{21} = g_{21} + jb_{21} = y_f = \left.\frac{\Delta i_2}{\Delta e_i}\right|_{e_2=0}$$

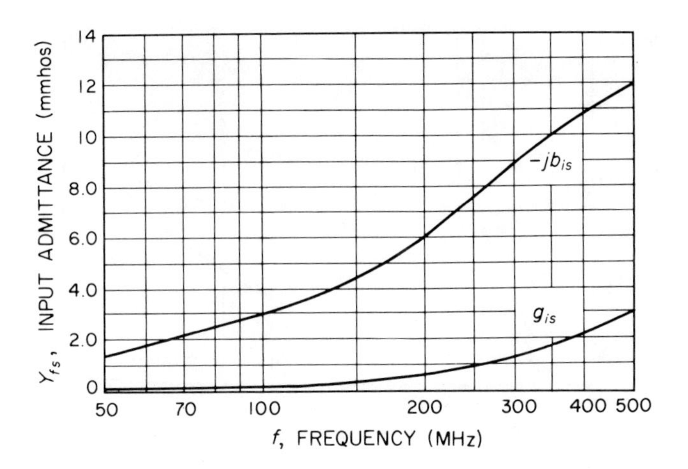

Fig. 2-106 Input admittance, $Y_{is} = g_{is} + jb_{is}$, of MFE3007 (Courtesy of Motorola Inc., Semiconductor Products Division)

Note that the data sheet of Fig. 2-13 shows y_{fs} or y_{21} at 1 kHz, and shows Re (y_{fs}), or the real part of y_{fs}, at 100 MHz. The minimum values are essentially the same. No maximum values for Re (y_{fs}) are given since they are unrealistic (due to gate-drain capacitive reactance).

Figure 2-107 shows more accurate and complete forward transadmittance curves for a typical FET. Note that the real and imaginary parts actually cross over at about 180 MHz. Also, the real part becomes a negative quantity at about 400 MHz.

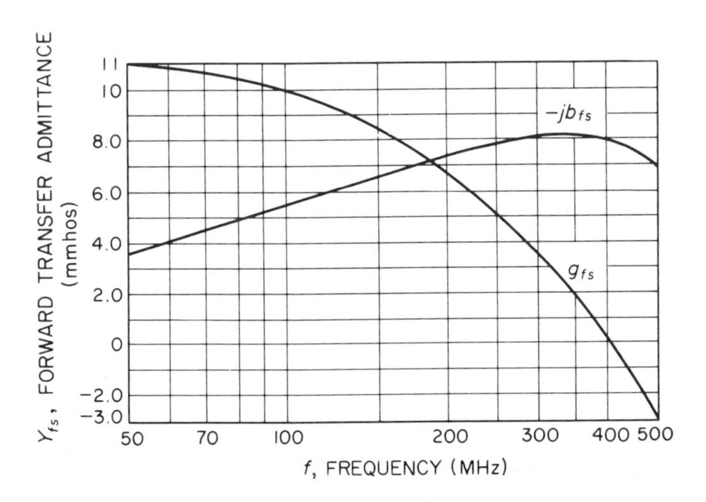

Fig. 2-107 Forward transfer admittance, $y_{fs} = g_{fs} + jb_{fs}$ of MFE3007 (Courtesy of Motorola Inc., Semiconductor Products Division)

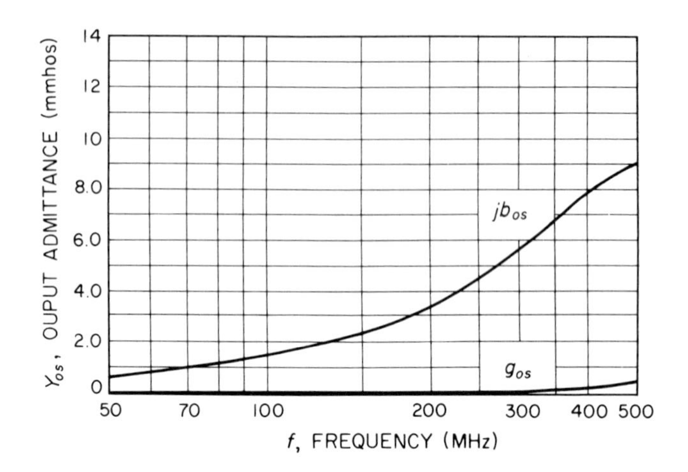

Fig. 2-108 Output admittance, $y_{os} = g_{os} + jb_{os}$, of MFE3007 (Courtesy of Motorola Inc., Semiconductor Products Division)

Output admittance, with Y_S = infinity (short circuit), is expressed as

$$y_{os} = y_{22} = g_{22} + jb_{22} = y_o = \left.\frac{\Delta i_2}{\Delta e_2}\right|_{e_1=0}$$

Note that the data sheet of Fig. 2-13 shows y_{os} at 1 kHz, but at no higher frequencies. Figure 2-108 shows more accurate and complete output admittance curves. The real part is negligible over the entire frequency range.

Reverse transadmittance, with Y_S = infinity (short circuit), is expressed as

$$y_{rs} = y_{12} = g_{12} + jb_{12} = y_r = \left.\frac{\Delta i_1}{\Delta e_2}\right|_{e_1=0}$$

Note that the data sheet of Fig. 2-13 does not show y_{rs} or y_{12} at any frequency. However, reverse transfer capacitance C_{rss} is given as 2 pF maximum. If we assume that the reverse transadmittance (or reverse transfer admittance, as it is sometimes called) is entirely (or mostly) capacitive (jb_{12}), then the reverse transadmittance can be found when C_{rss} is multiplied by $6.28F$ (F = frequency in Hz). This assumption is generally accurate in the case of y_{rs} or y_{12}, as shown in Fig. 2-109. Note that the real part of y_{rs} (or g_{rs}) is zero across the entire frequency range. Therefore, when the term Re (y_{12}) or Re (y_{rs}) appears, it can be considered as zero for all practical design purposes.

y-Parameter measurement. As can be seen thus far, y-parameter information is not always available, or not in a convenient form. In practical design, it may be necessary to measure the y-parameters, using laboratory equipment.

Note that all y parameters are based on ratios of input/output current to input/output voltage. For example, y_{fs} is the ratio of output current to input voltage.

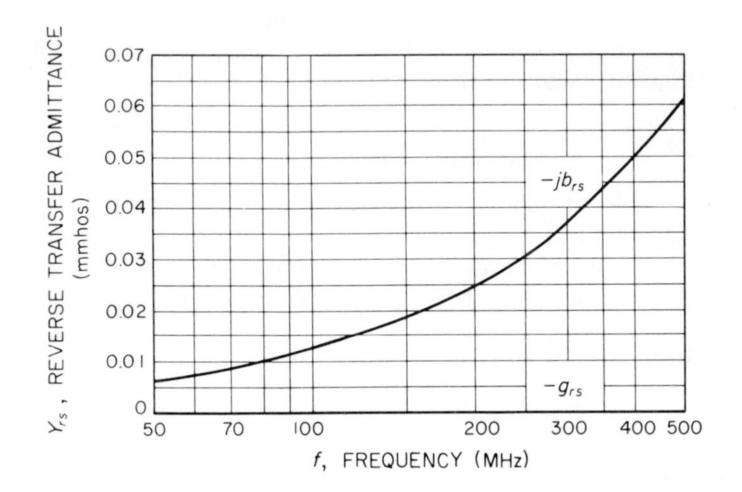

Fig. 2-109 Reverse transfer admittance, $y_{rs} = g_{rs} + jb_{rs}$, of MFE3007 (Courtesy of Motorola Inc., Semiconductor Products Division)

Y_{fs} and Y_{os} can be measured by using signal generators, voltmeters, and simple circuits. Likewise, y_{is} and y_{rs} can be found by measuring c_{iss} and c_{rss} (using a simple capacitance meter), and then calculating the y_{is} and y_{rs} based on frequency of interest. (All of these procedures are described in the author's *Practical Semiconductor Databook for Electronic Engineers and Technicians.*)

However, more accurate results will be obtained if precision laboratory equipment is used. All four y-parameters can be measured on a General Radio Transfer Function and Immittance Bridge. A possible exception is y_{rs} which is typically very small in relation to the other parameters. In the case of y_{rs}, it is often more practical to measure c_{rss} and multiply by $6.28f$.

The main concern in measuring y-parameters, from a practical design standpoint, is that the measurements are made under conditions simulating those of the final circuit. For example, the supply or drain-source voltage, gate-source voltage, bias (if any), and operating frequency should be identical (or close) to the final circuit. Otherwise, the tests can be misleading.

2-26-2 Stability Factors

There are two factors used to determine the potential stability (or instability) of FETs in RF amplifiers. (Note that these same factors are used with other devices such as bipolar transistor RF amplifiers and IC amplifiers.) One factor is known as the Linvill C factor; the other is the Stern k factor. Both factors are calculated from equations requiring y-parameter information (to be taken from data sheets or by actual measurement at the frequency of interest). The main difference between the two is that the Linvill C factor assumes the FET is not connected to a load, while the Stern k factor includes the effect of a given or specific load.

The Linvill C factor is calculated from

$$C = \frac{y_{12}y_{21}}{2g_{11}g_{22} - \mathrm{Re}(y_{12}y_{21})}$$

If C is less than 1, the FET is unconditionally stable; that is, using a conventional (unmodified) circuit, no combination of load and source admittances can be found which will cause oscillation. If C is greater than 1, the FET is potentially unstable; that is, certain combinations of load and source admittances will cause oscillation.

The Stern k factor is calculated from

$$k = \frac{2(g_{11} + G_S)(g_{22} + G_L)}{y_{12}y_{21} + \mathrm{Re}(y_{12}y_{21})}$$

where G_S and G_L are source and load conductances, respectively (G_S = 1/source resistance; G_L = 1/load resistance).

If k is greater than 1, the amplifier circuit is stable (opposite from Linvill). If k is less than 1, the amplifier is unstable. In practical design, it is recommended that a k factor of 3 or 4 be used, rather than 1, to provide a margin of safety. This will accommodate parameter and component variations (particularly with regards to band-pass response).

Note that both equations are fairly complex, and require considerable time for their solution (unless computer-aided design techniques are used).

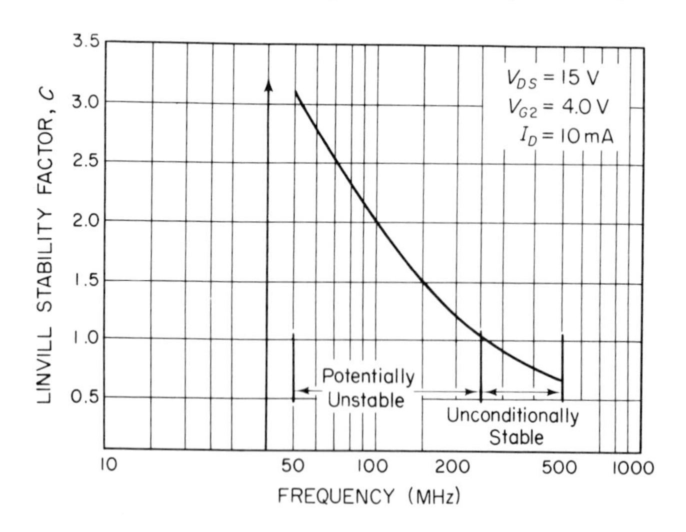

Fig. 2-110 Linvill stability factor, C, for the MFE3007 between 50 and 500 MHz (Courtesy of Motorola Inc., Semiconductor Products Division)

Some manufacturers provide alternate solutions to the stability and load-matching problems, usually in the form of a data-sheet graph. Such a graph is shown in Fig. 2-110, which is a Linvill C factor chart for a typical FET. Note that the FET is unconditionally stable at frequencies above 250 MHz, but potentially unstable at frequencies below 250 MHz. At frequencies below about 50 MHz, the FET becomes highly unstable.

2-26-3 Solutions to Stability Problem

There are two basic solutions to the problem of unstable RF amplifiers. First, the amplifier can be neutralized. That is, part of the output can be fed back (shifted in phase) to the input so as to cancel oscillation. This solution requires extra components, and creates a problem when frequency is changed. The other solution is to introduce some mismatch into either the source or load tuning networks. This solution requires no extra components, but does produce a reduction in gain.

A comparison of the two methods is shown in Fig. 2-111. The higher gain curve represents the unilateralized (or neutralized) operation. The lower gain curve represents the circuit power gain, when the Stern k factor is 3. Assume that the frequency of interest is 100 MHz. If the amplifier is matched directly to the load and source, without regard to stability or with neutralization to produce stability, the top curve applies, and the power gain is about 38 dB. If the amplifier is matched to a load and source where the Stern k

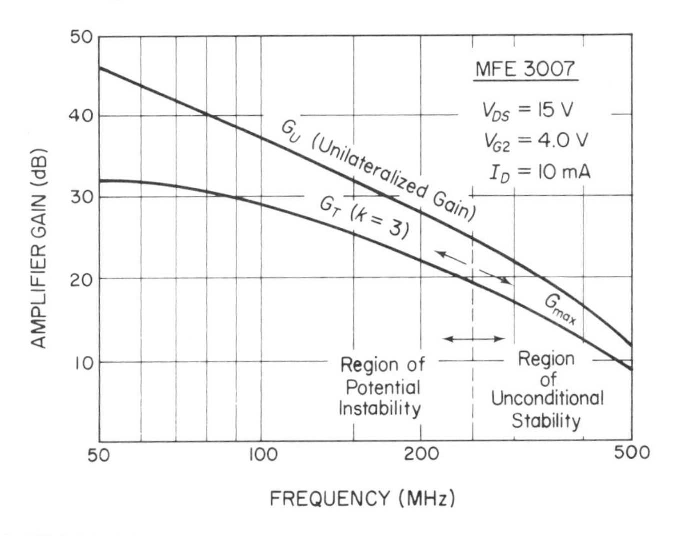

Fig. 2-111 Amplifier gain characteristics in common-source configuration (Courtesy of Motorola Inc., Semiconductor Products Division)

factor is 3 (probably resulting in a mismatch with the actual load and source), the lower curve applies, and the power gain is about 29 dB.

The upper curve of Fig. 2-111 is found by the *general power gain equation*:

$$G_P = \frac{\text{Power delivered to load}}{\text{Power delivered to input}}$$

$$= \frac{|Y_{21}|^2 G_L}{(Y_L + y_{22})^2 \, \mathrm{Re}\left(y_{11} - \dfrac{y_{12}y_{21}}{y_{22} + Y_L}\right)}$$

The general power gain equation applies to circuits with no external feedback, and to circuits which have external feedback (neutralization), provided the composite y parameters of both FET and feedback network are substituted for the FET y parameters in the equation.

The lower curve of Fig. 2-111 is found by the *transducer gain expression*:

$$G_T = \frac{\text{Power delivered to load}}{\text{Maximum power available from source}}$$

$$= \frac{4G_S G_L |y_{21}|^2}{[(y_{11} + Y_S)(y_{22} + Y_L) - y_{12}y_{21}]^2}$$

The transducer gain expression includes input mismatch. The lower curve of Fig. 2-111 assumes that the mismatch is such that a Stern k factor of 3 results. That is, the circuit tuning networks are adjusted for admittances that produce a Stern k factor of 3. The transducer gain expression considers the input and output networks as part of the source and load.

With either gain expression, the input and output admittances of the FET are modified by the load and source admittances.

The input admittance of the FET is given by:

$$Y_{\mathrm{IN}} = y_{is}\,(\text{or } y_{11}) - \frac{y_{12}y_{21}}{y_{22} + Y_L}$$

The output admittance of the FET is given by:

$$Y_{\mathrm{OUT}} = y_{os}(\text{or } y_{22}) - \frac{y_{12}y_{21}}{y_{11} + Y_S}$$

At low frequencies, the second term in the input and output admittance equations is not particularly significant. At VHF, the second term makes a very significant contribution to the input and output admittances.

The imaginary parts of Y_S and Y_L (B_S and B_L, respectively) must be known before values can be calculated for power gain, transducer gain, input admittance, and output admittance. Except for some very special cases, exact solutions for B_S and B_L consist of time-consuming complex algebraic manipulations.

As fairly good simplifying approximations for the equations, let $B_S \approx -b_{11}$ and $B_L \approx -b_{22}$ so that:

general power gain expression

$$G_P \approx \frac{|y_{21}|^2 G_L}{(G_L + g_{22})^2 \operatorname{Re}\left(y_{11} - \dfrac{y_{12}y_{21}}{g_{22} + G_L}\right)}$$

transducer gain expression

$$G_T \approx \frac{4G_S G_L |y_{21}|^2}{[(g_{11} + G_S)(g_{22} + G_L) - y_{12}y_{21}]^2}$$

input admittance

$$Y_{\text{IN}} \approx y_{11} - \frac{y_{12}y_{21}}{g_{22} + G_L}$$

output admittance

$$Y_{\text{OUT}} \approx y_{22} - \frac{y_{12}y_{21}}{g_{11} + G_S}$$

2-26-4 The Stern Solution

A stable design with a potentially unstable FET is possible without external feedback (neutralization) by proper choice of source and load admittances. This can be seen by inspection of the Stern k factor equation; G_S and G_L can be made large enough to yield a stable circuit, regardless of the degree of potential instability. Using this approach, a circuit stability factor k (typically $k = 3$) is selected, and the Stern k-factor equation is used to arrive at values of G_S and G_L which will provide the desired k. Of course, the actual G of the source and load can not be changed. Instead, the input and output tuning circuits are designed as if the actual G values were changed. This results in a mismatch, and a reduction in power gain, but does produce the desired degree of stability.

To get a particular circuit stability factor, the designer may choose any of the following combinations of matching or mismatching of G_S and G_L to the FET input and output conductances, respectively:

G_S matched and G_L mismatched

G_L matched and G_S mismatched

both G_S and G_L mismatched

Often a decision on which combination to use will be dictated by other performance requirements or practical considerations.

Once G_S and G_L have been chosen, the remainder of the design may be completed by using the relationships which apply to the amplifier without feedback. Power gain and input and output admittances may be computed by using the appropriate equations (Sec. 2-25-3).

Simplified Stern approach. Although the above procedure may be adequate in many cases, a more systematic method of source and load admittance determination is desirable for designs which demand maximum power gain per degree of circuit stability. Stern has analyzed this problem and developed equations for computing the optimum G_S, G_L, B_S, and B_L for a particular circuit stability factor (Stern k factor). Unfortunately, these equations are very complex and quite tedious if they must be done frequently. The complete Stern solution is best solved by computer. As a matter of interest, a program has been written in BASIC to provide essential information for FETs used as RF amplifiers. A second program has been written to include the effects of a specific source and load. This second program permits the designer to experiment with theoretical "breadboard" circuits in a matter of seconds. Other programs perform parameter conversions and the network synthesis for FET amplifier design.

When a Stern solution must be obtained without the aid of a computer, it is best to use one of the many short-cuts that have been developed over the years. The following short-cut is by far the simplest and most widely accepted, yet provides an accuracy close to that of the computer solutions.

1. Let $B_S \approx -b_{11}$ and $B_L \approx -b_{22}$, as in the case of the Sec. 2-25-3 equations. This method permits the designer to closely approximate the exact Stern solution for Y_S and Y_L, while avoiding that portion of the computations which is the most complex and time consuming. Further, the circuit can be designed with tuning adjustments for varying B_S and B_L, thereby creating the possibility of experimentally achieving the true B_S and B_L for maximum gain as accurately as if all the Stern equations had been solved.

2. Mismatch G_S to g_{11} and G_L to g_{22} by an *equal ratio*. That is, find a ratio that produces the desired Stern k factor, and then mismatch G_S to g_{11} (and G_L to g_{22}). For example, if the ratio is 4-to-1, make G_s four times the value of g_{11} (and G_L four times the value of G_{22}).

If the mismatch ratio, R, is defined as

$$R = \frac{G_L}{g_{22}} = \frac{G_S}{g_{11}}$$

then R may be computed for any particular circuit stability (k) factor by using the equation:

$$R = \left(\sqrt{k\left[\frac{|y_{21}y_{12}| + \mathrm{Re}(y_{12}y_{21})}{2g_{11}g_{22}}\right]}\right) - 1$$

This short-cut method may be advantageous if source and load admittances and power gains for several different values of k are desired. Once the

R for a particular k has been determined, the R for any other k may be quickly found from the equation:

$$R = \frac{(1 + R_1)^2}{(1 + R_2)^2} = \frac{k_1}{k_2}$$

where R_1 and R_2 are values of R corresponding to k_1 and k_2, respectively.

The Stern solution with data-sheet graphs. It is obvious that the Stern solution, even the short-cut method, is somewhat complex. For this reason, some manufacturers have produced data-sheet graphs that show optimum source and load admittances for a particular FET over a wide range of frequencies. Examples of these graphs are shown in Figs. 2-112 and 2-113. Figure 2-112 shows both the real (G_S) and imaginary (B_S) values that will produce maximum gain, but with a stability (Stern k) factor of 3, at frequencies from 50 to 500 MHz. Figure 2-113 shows corresponding information for G_L and B_L. To use these illustrations, simply select the desired frequency, and note where the corresponding G and B curves cross the frequency line. For example, assuming a frequency of 100 MHz, $Y_L = 0.35 - j2.1$ mmhos, and $Y_S = 1.30 - j4.4$ mmhos.

If the tuning circuits are designed to match these admittances, rather than the actual admittances of the source and load, the circuit will be stable. Of course, the gain will be reduced. Use the transducer gain expression G_T of Sec. 2-25-3 to find the resultant power gain.

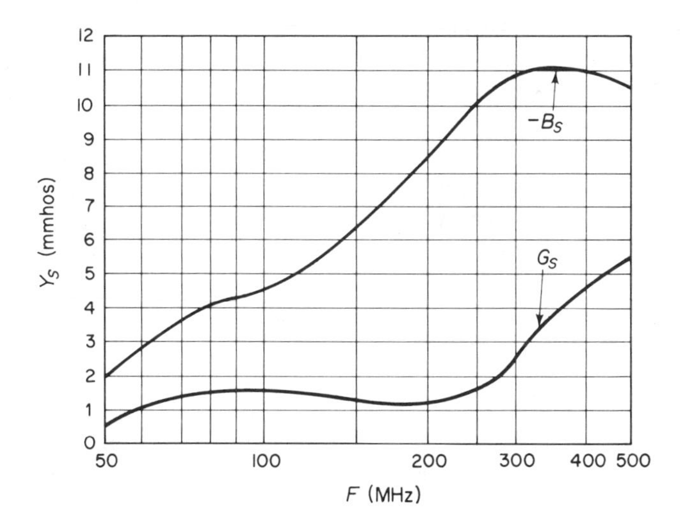

Fig. 2-112 Optimum source admittance, $Y_S = G_S + jB_S$ (Courtesy of Motorola Inc., Semiconductor Products Division)

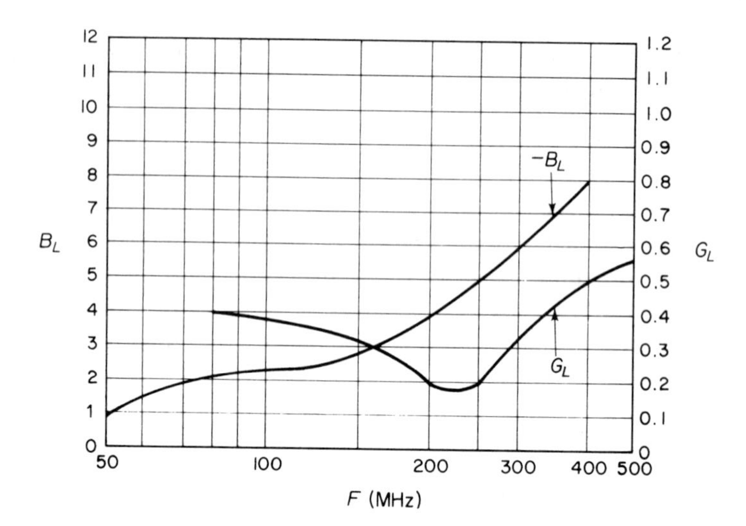

Fig. 2-113 Optimum load admittance, $Y_L = G_L + jB_L$ (Courtesy of Motorola Inc., Semiconductor Products Division)

2-26-5 *Systematic* FET RF *Amplifier Design Procedure*

A review of the two-part (y parameter) network design method may be helpful at this point. Basically, the steps are:

1. Determine the potential instability of the FET. This involves extracting the y-parameters of the FET from the data sheet, or determining the y-parameters from actual test, as described in Sec. 2-25-1. Next, plug the y-parameters into the Linvill C and/or Stern k equations to find potential stability or instability. Use the Linvill C factor where source and load impedances are not involved (or known). Use the Stern k factor when load and source impedances are known. As a practical matter, it is usually more convenient to go directly to the Stern k factor, since this serves as a starting point if the circuit must be modified to produce stability.

2. If the FET is not unconditionally stable, decide on a course of action to insure circuit stability. Usually, this involves going to neutralization, or mismatching input/output tuning circuits. Mismatching is, by far, the most popular course of action. If the FET is unconditionally stable, without neutralization or mismatch, the design can proceed to fulfill other objections without fear of oscillation. Under these circumstances, the usual object is to get maximum gain by matching the tuning circuits to the actual source and load.

3. Determine source and load admittances. Source and load admittance determination is dependent upon gain and stability considerations, together with practical circuit limitations. If the FET is potentially unstable at the frequency of interest and with actual source and load impedances, then a source and load which will guarantee a certain degree of circuit stability must be used. This involves the Stern solution described in Sec. 2-25-4. If optimum source and load impedances are given by a manufacturer's data sheet, use these as a first choice. As a second choice, use computer-aided design techniques to get a Stern solution for the desired stability and gain. If neither of these is available, use the short-cut Stern technique of Sec. 2-25-4. Note that it is a good idea to check circuit stability (Stern k) factor, even when an unconditionally stable FET has been found by the Linvill C factor. A FET may be stable without a load, or with certain loads, but not stable with some specific load.

Once the optimum source and load admittances have been selected, verify that the required gain will be available. In practical terms, it is possible to mismatch almost any FET RF amplifier sufficiently to produce a stable circuit. However, the resultant power gain may be below that required. In that case, a different FET must be used (or a lower gain accepted).

4. Design appropriate networks (input/output tuning circuits) to provide the desired (or selected) source and load admittances. First, the networks must be resonant at the desired frequency. (That is, inductive and capacitive reactance must be equal at the selected frequency.) Second, the network must match the FET to the load and source. Sometimes it will be difficult to achieve a desired source and load due to tuning range limitations, excess network losses, component limitations, etc. In such cases, the source and load admittances will be a compromise between desired performance and practical limitations. Generally, this involves a sacrifice of gain to achieve stability.

The remainder of this section is devoted to the step-by-step design procedures for two FET RF amplifier circuits.

2-26-6 *Design Example of 100-MHz Common-Source* FET RF *Amplifier*

Assume that the circuit of Fig. 2-114 is to be operated at 100 MHz. The source and load impedances are both 50 ohms. The characteristics of the FET are given in Figs. 2-106 through 2-112. The problem is to find the optimum values of C_1 through C_4, as well as L_1 and L_2.

From Fig. 2-110, the Linvill stability factor C is seen to be 2.0. Therefore, mismatching or neutralization is necessary to prevent oscillation. Mismatching will be used in this example. Figure 2-111 shows that for a circuit

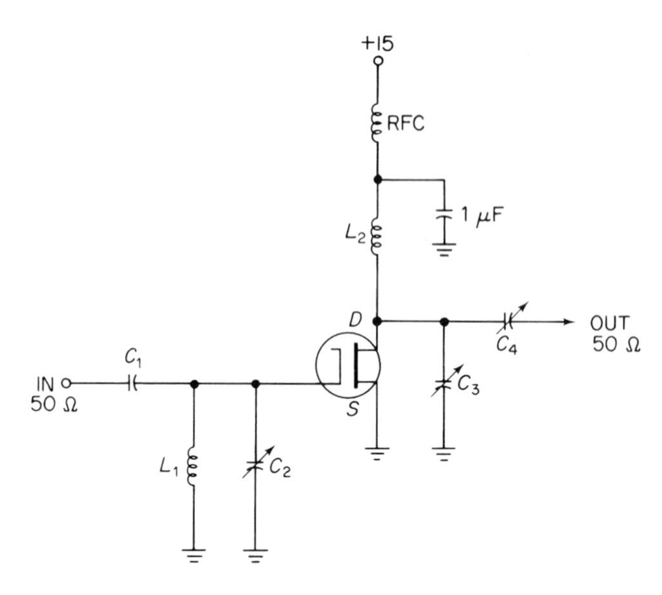

Fig. 2-114 Common-source FET RF amplifier (Courtesy of Motorola Inc., Semiconductor Products Division)

stability (Stern k) factor of 3.0, the transducer gain will be 29 dB. The load and source admittances for the required mismatch and gain (found in Figs. 2-112 and 2-113) are:

$$Y_L = 0.35 - j2.1 \text{ mmhos}$$
$$Y_S = 1.30 - j4.4 \text{ mmhos}$$

At 100 MHz, y parameters for the FET are:

$$y_{is} = y_{11} = 0.15 + j3.0 \text{ (Fig. 2-106)}$$
$$y_{fs} = y_{21} = 10 - j5.5 \text{ (Fig. 2-107)}$$
$$y_{rs} = y_{12} = 0 - j0.012 \text{ (Fig. 2-109)}$$
$$y_{os} = y_{22} = 0.04 + j1.7 \text{ (Fig. 2-108)}$$

Assuming the load and source impedances are both 50 ohms, networks can be designed to match the FET to the load and source. The calculations are more easily performed with impedances rather than admittances. The procedure will first be discussed for the output matching.

The 50-ohm load impedance must be transformed to the optimum load for the FET ($Y_L = 0.35 - j2.1$). This transformation can be performed by the network shown in Fig. 2-115a. In effect, R_L is in series with C_4. The 50 ohms must be transformed to

$$R_L = \frac{1}{G_L} = \frac{1}{0.35 \times 10^{-3}} \approx 2.86 \text{ k}\Omega$$

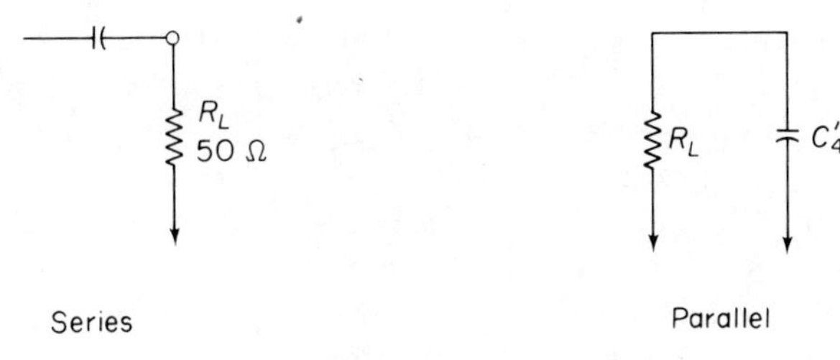

(a) OUTPUT IMPEDANCE TRANSFORMATION

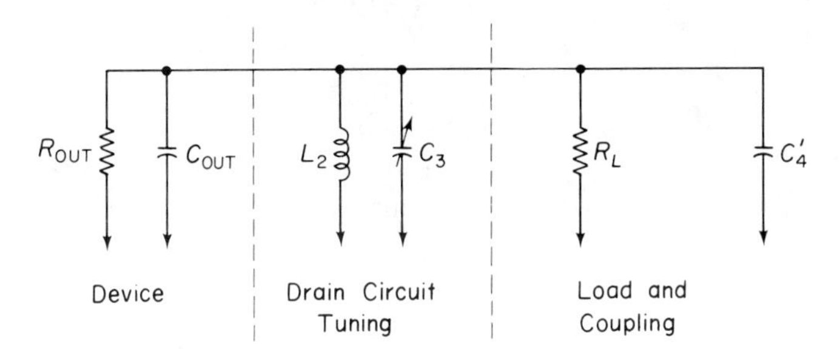

(b) EQUIVALENT OUTPUT CIRCUIT

Fig. 2-115 Impedance transformation and equivalent output circuit of common-source FET RF amplifer (Courtesy of Motorola Inc., Semiconductor Products Division)

The series capacitive reactance required for ths matching can be found by

$$X_{C_4} = X_{\text{series}} = R_S \sqrt{\frac{R_P}{R_S} - 1}$$

where R_P is the parallel resistance and R_S is the series resistance.

$$X_{C_4} = 50\sqrt{\frac{2.86 \times 10^3}{50} - 1} \approx 372 \text{ ohms}$$

The capacitance that provides this reactance at 100 MHz is

$$C_4 = \frac{1}{6.28FX_{C_4}} = \frac{1}{6.28(10^8)(372)} \approx 4.3 \text{ pF}$$

The parallel equivalent of this capacitance is needed for determining the bandwidth and resonance later in the design:

$$X'_{C_4} = X_{\text{Parallel}} = X_S\left[1 + \left(\frac{R_S}{X_S}\right)^2\right]$$

$$= 372\left[1 + \left(\frac{50}{372}\right)^2\right] \approx 378 \text{ ohms}$$

and the equivalent parallel capacitance is therefore

$$C'_4 \approx 4.2 \text{ pF}$$

An equivalent circuit for the output tank after transformation of the load is shown in Fig. 2-115b. Since the resistance across the output circuit is fixed by the parallel combination of R_{OUT} and R_L (after transformation), the desired bandwidth of the output tank will be determined by C_3.

Note that the output admittance (Y_{OUT}) of the FET will not equal y_{os} under most conditions. Only when the input is terminated in a short circuit, or the feedback admittance is zero, does Y_{OUT} equal y_{os}. When y_{rs} is not zero and the input is terminated with a practical source admittance, the true output admittance is found from the following (or from actual test):

$$Y_{OUT} = Y_{os} - \frac{y_{fs}Y_{rs}}{y_{is} + Y_S}$$

$$= 0.04 + j1.7 - \frac{(10 - j5.5)(0 - j0.012)}{(0.15 + j3.0) + (1.3 - j4.4)}$$

$$= -0.066 + j2.05 \text{ mmhos}$$

therefore,

$$R_{OUT} = \frac{1}{G_{OUT}} = \frac{1}{-0.066 \times 10^{-3}} = -15.2 \text{ k}\Omega$$

$$C_{OUT} = \frac{B_{OUT}}{6.28F} = \frac{2.05 \times 10^{-3}}{6.28(10^8)} \approx 3.2 \text{ pF}$$

The negative output impedance indicates the instability of the unloaded amplifier.

Now the total impedance across the output tank can be calculated:

$$R_T = \frac{1}{G_{OUT} + G_L}$$

$$= \frac{1}{-0.066 \times 10^{-3} + 0.35 \times 10^{-3}} \approx 3.52 \text{ k}\Omega$$

Since the output impedance is several times higher than the input impedance of the FET, amplifier bandwidth is primarily dependent upon output-loaded Q. For a bandwidth of 5 MHz (3-dB points),

$$C_T = \frac{1}{6.28R_T(\text{BW})}$$

$$= \frac{1}{6.28(3.52 \times 10^3)(5 \times 10^6)} \approx 9\text{pF}$$

hence,

$$C_3 = C_T - C_{OUT} - C'_4 = 9.0 - 3.2 - 4.2 = 1.6 \text{ pF}$$

The output inductance that resonates with C_T at 100 MHz is 280 nH. This completes the output circuit design.

Input calculations performed in a similar manner yield these results:

$$Y_S = 1.30 - j4.4 \text{ mmhos}$$

$$X_{C_1} = 190 \text{ ohms; therefore, } C_1 = 8.4 \text{ pF}$$

$$X'_{C_1} = 203 \text{ ohms; therefore, } C'_1 = 7.8 \text{ pF}$$

$$Y_{IN} = y_{is} - \frac{y_{fs}y_{rs}}{y_{os} + Y_L} = -0.25 + j4.52 \text{ mmhos;}$$

therefore, $R_{IN} = -4\ k\Omega,\ C_{IN} = 7.2$ pF

$$R_T = 950 \text{ ohms}$$

The bandwidth of the input tuned circuit is chosen to be 10 MHz. Hence,

$$C_T = 17 \text{ pF; therefore, } L_1 = 150 \text{ nH}$$

$$C_2 = 17 - 7.2 - 7.8 = 2 \text{ pF}$$

This completes design of the tuned circuits. It is important that the circuit be well bypassed to ground at the signal frequency, since only a small impedance to ground may cause instability or loss in gain. The bypass capacitor should be such that the reactance is about 1 to 2 ohms at the operating frequency. A 1-μF capacitor will provide less than 1-ohm reactance at 100 MHz.

2-26-7 Design Example of 87.5–108.5 MHz Common-Gate FET RF *Amplifier*

Assume that the circuit of Fig. 2-116 is to be tuned across the range from 87.5 to 108.5 (broadcast FM) MHz. An equivalent circuit of the RF stage to be designed is given in Fig. 2-117. Coil turns are defined and high-frequency conductances are shown in Fig. 2-117 where:

G_S^* = stage driving source conductance
G_{IN}^* = stage input conductance
G_S = FET driving source conductance
G_{IN} = FET input conductance with a finite load
G_{OUT} = FET output conductance with a finite driving source conductance
G_L = FET load conductance
G_L^* = stage load conductance

The initial arbitrary chosen assumptions and design goals for this example are:

1. $G_S^* = 3.33$ mmho (300-ohm antenna)
2. $G_{IN}^* = G_S^*$ (for minimum antenna VSWR)
3. $G_L^* = 4.0$ mmho (typical input conductance of a bipolar transistor mixer)
4. Selectivity at 21.4 MHz above center frequency ≥ 45 dB (needed for image rejection ≥ 45 dB)
5. Tuning range = 87.5 to 108.5 MHz
6. Tune with standard variable capacitor: $C_{1A} = C_{1B} = 6$—21 pF ($\Delta C = 15$ pF)
7. Coils must have practical tapping ratios, wire sizes, and shape factors
8. Gain control will be reverse AGC
9. Observing the above restrictions, achieve high power gain and low noise figure, using an *N*-channel silicon TIS34 JFET.

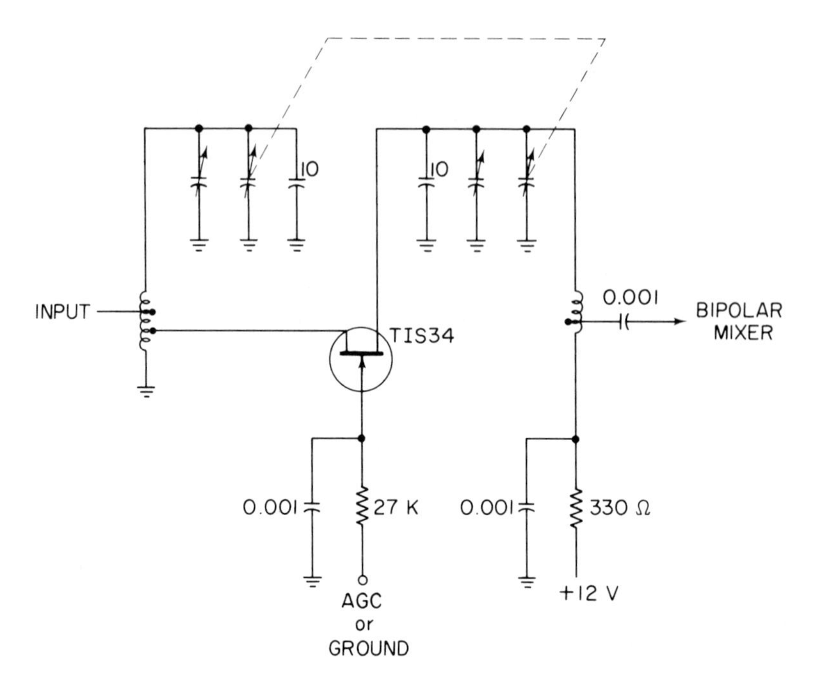

Fig. 2-116 Common-gate FET RF amplifier (Courtesy, Texas Instruments Incorporated)

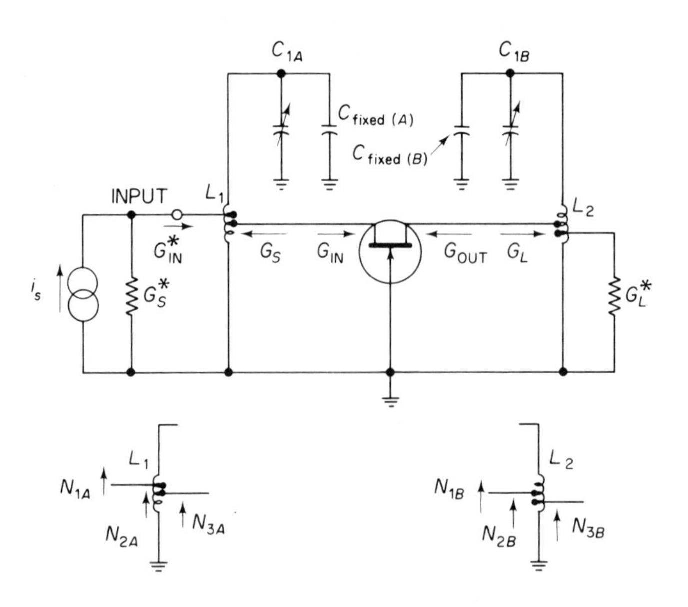

Fig. 2-117 A-C equivalent circuit of common-gate FET RF amplifier (Courtesy, Texas Instruments Incorporated)

Given a ΔC for the tuning capacitor and a desired tuning range, the input and output coil inductances and fixed capacitances are required to be

$$L_1 = \frac{1}{(6.28\,F_2)^2\left[\dfrac{\Delta C_{1A}}{\left(\dfrac{F_2}{F_1}\right)^2 - 1}\right]}$$

$$L_2 = \frac{1}{(6.28\,F_2)^2\left[\dfrac{\Delta C_{1B}}{\left(\dfrac{F_2}{F_1}\right)^2 - 1}\right]}$$

$$C_{\text{fixed}(A)} = \frac{\Delta C_{1A}}{\left(\dfrac{F_2}{F_1}\right)^2 - 1} - C_{1A(\text{min})}$$

$$C_{\text{fixed}(B)} = \frac{\Delta C_{1B}}{\left(\dfrac{F_2}{F_1}\right)^2 - 1} - C_{1B(\text{min})}$$

where F_1 = lowest signal frequency and F_2 = highest signal frequency.

For this particular example, $C_{1A(\text{min})} = C_{1B(\text{min})}$, $\Delta C_{1A} = \Delta C_{1B}$. $F_1 =$ 87.5 MHz and $F_2 =$ 108.5 MHz, so that:

$$L_1 = L_2 = 0.077\ \mu\text{H};$$

$$C_{\text{fixed}(A)} = C_{\text{fixed}(B)} = 22\ \text{pF}$$

The preceding values for capacitances $C_{\text{fixed}(A)}$ and $C_{\text{fixed}(B)}$ include trimmer capacitances, stray capacitances, and reflected FET input and output capacitances, in addition to any actual fixed capacitors placed across the input or output tanks. With a typical trimmer setting of about 8 pF, and allowing a total of about 4 pF for stray capacitance and FET reflected capacitance, a 10-pF capacitor can be placed across the output tank.

The remainder of the design is carried out at signal frequency $F_0 = 100$ MHz where:

$$X_C = X_L = 6.28 F_0 L = 48.4\ \text{ohms}$$

Based on the discussion of biasing in Secs. 2-5 and 2-8, the values of the drain and gate resistances shown in Fig. 2-116 result in an operating point at I_{DSS}. Typical *common-source* y-parameters for the TIS34 at I_{DSS} are:

$$y_{is} = y_{11} = 0.10 + j3.00\ \text{mmho}$$

$$y_{fs} = y_{21} = 4.80 - j1.20\ \text{mmho}$$

$$y_{rs} = y_{12} = 0 - j0.95\ \text{mmho}$$

$$y_{os} = y_{22} = 0.05 + j1.05\ \text{mmho}$$

Since the circuit is common-gate, the common-source parameters must be converted to common-gate as follows:

$$y_{ig} = y_{is} + y_{fs} + y_{rs} + y_{os} = \quad 4.95 + j1.9 \text{ mmho}$$

$$y_{fg} = -(y_{fs} + y_{os}) = -4.85 + j0.15 \text{ mmho}$$

$$y_{rg} = -(y_{rs} + y_{os}) = -0.05 - j0.1 \text{ mmho}$$

$$y_{og} = y_{os} = \quad 0.05 + j1.05 \text{ mmho}$$

Using these common-gate y-parameters, and the Linvill C factor discussed in Sec. 2-25-2, the stability is

$$C = \frac{y_{12}y_{21}}{2g_{11}g_{22} - \text{Re}(y_{12}y_{21})} = \frac{0.545}{2(4.95)(0.05) - 0.258} = 2.3$$

Since C is greater than 1, the FET is potentially unstable in the common-gate configuration.

Mismatching is to be used for the required stability. A Stern k factor of 4 is chosen. Using the short-cut Stern solution described in Sec. 2-25-4 (mismatching G_S and G_L to g_{11} and g_{22} by an equal ratio R), the value of R is

$$R = \frac{4(0.545 + 0.258)}{2(4.95)(0.05)} \approx 1.56$$

Using a value of 1.56 for R, and the equation

$$1.56 = \frac{G_S}{g_{11}} = \frac{G_L}{g_{22}}$$

then
$$G_S = (1.56)(4.95)(10^{-3}) = 7.72 \text{ mmho}$$

$$\left(R_S = \frac{1}{G_S} = 130 \text{ ohms}\right)$$

$$G_L = (1.56)(0.05)(10^{-3}) = 0.078 \text{ mmho}$$

$$\left(R_L = \frac{1}{G_L} = 12.8 \text{ k}\Omega\right)$$

If it were practical to realize such a small G_L (or a high R_L), the power gain and transducer gain from the equations of Sec. 2-25-3 would be

$$G_P \approx \frac{(y_{21})^2 G_L}{(G_L + g_{22})^2 \text{ Re}\left(y_{11} - \dfrac{y_{12}y_{21}}{g_{22} + G_L}\right)} = 15.8 \text{ dB}$$

$$G_T \approx \frac{4G_S G_L (y_{21})^2}{[(g_{11} + G_S)(g_{22} + G_L) - y_{12}y_{21}]^2} = 14.3 \text{ dB}$$

Unfortunately, G_L is too small to be conveniently realized. The above calculations are informative, however, in that they suggest that, *for the common-gate configuration*, the FET should work into as high an impedance load as is practical. This says that the drain terminal of the FET will go to the top of the output tank ($N_{1B}/N_{2B} = 1.0$), that the actual value of R_L^* will be tapped as low on the output tank as is practical, and that the highest practical value for the unloaded coil Q should be chosen.

Given as a guide for the selection of Q_U, Figs. 2-118, 2-119, and 2-120 represent data taken at 100 MHz on hand-wound coils using a Boonton Radio Co. 190-A high-frequency Q meter.

Figure 2-118 shows how Q_U is a function of the internal diameter of a coil, when no core is used and the wire gauge and inductance of the coil are held constant. Assuming no differences in the proximity of the coils to a metal chassis due to different coil sizes, the larger-diameter coils have the higher Q_U.

Figure 2-119 illustrates how Q_U varies with wire gauge for a constant inductance and a constant internal diameter. The Q_U coils without cores are fairly sensitive to the wire gauge, but those using Carbonyl J and Carbonyl E cores are almost independent of the wire gauge.

In Fig. 2-120, Q_U is given as a function of the penetration of different types of cores into the coil. Curves are given for Carbonyl J, Carbonyl E, and aluminum cores, with the wire gauge, internal diameter, and inductance held constant.

The circuit of Fig. 2-116 is taken from an FM receiver front end. Practical layout and shielding requirements usually require that the FM front end be placed in its own small metal chassis. Because the coils are close to metal with

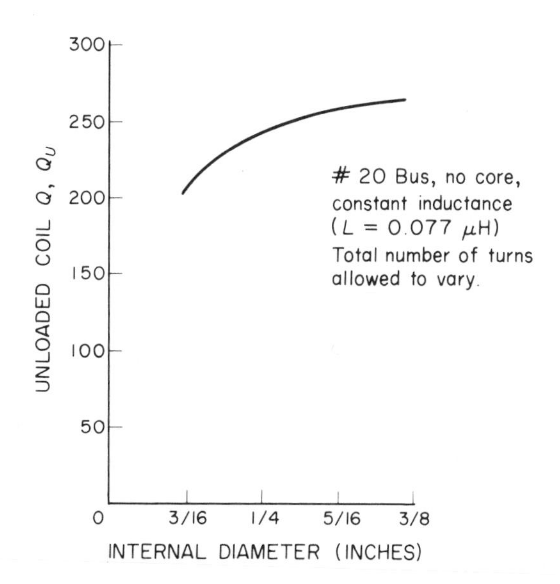

Fig. 2-118 Unloaded coil Q as a function of coil internal diameter (Courtesy, Texas Instruments Incorporated)

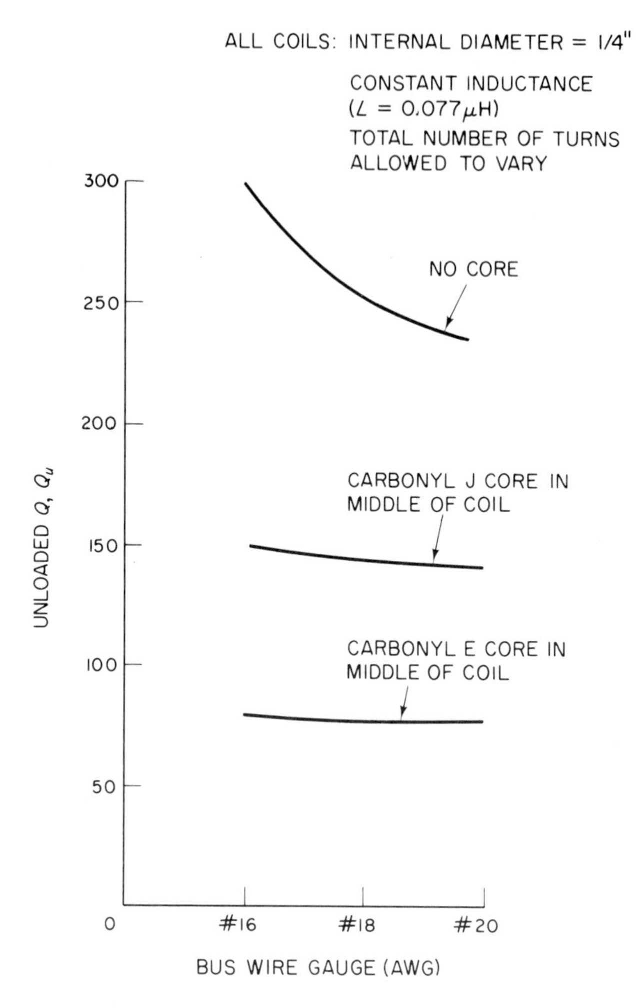

Fig. 2-119 Unloaded coil Q as a function of wire gauge (Courtesy, Texas Instruments Incorporated)

such an arrangement, the unloaded Q is reduced. Likewise, the presence of solder on the coils can reduce unloaded Q.

With these comments in mind, in selecting as high an impedance for the output circuit as practical, the following somewhat arbitrary limits are imposed:

$$Q_U = 200 \text{ minimum } (Q_U X_C = 200 \times 48.6 \approx 9.6 \text{ k}\Omega)$$

$$\frac{N_{1B}}{N_{3B}} = 5.0 \text{ maximum}$$

$$\frac{N_{1B}}{N_{2B}} = 1.0$$

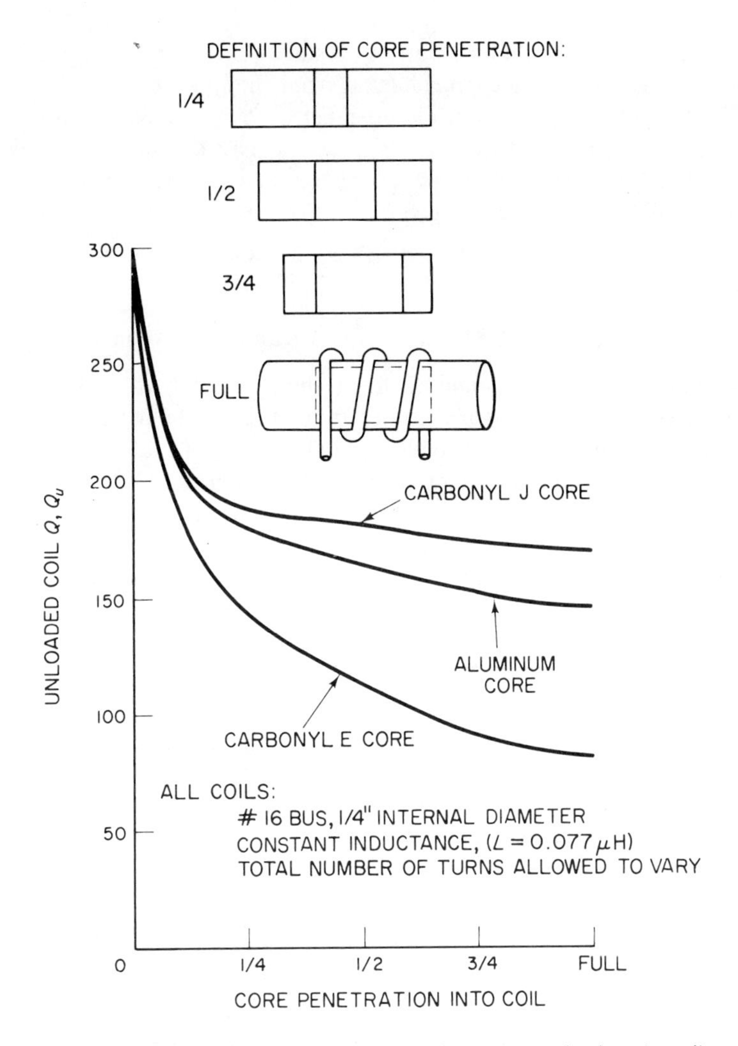

Fig. 2-120 Unloaded coil Q as a function of core penetration into the coil (Courtesy, Texas Instruments Incorporated)

These restrictions give:

$$G_{L(\text{min})} = \frac{1}{Q_U X_{C(\text{max})}} + \frac{G_L^*}{(N_{1B}/N_{3B(\text{min})})^2}$$

$$G_{L(\text{min})} = 0.304 \text{ mmho or } R_L \text{ max} = 3.29 \text{ k}\Omega$$

Temporarily assuming that the output resistance of the common-gate FET will not appreciably load the output tank, the loaded Q of the output circuit will be

$$Q_L \approx \frac{R_{L(\text{max})}}{X_C} = 68$$

It will be seen that the FET output resistance is quite high, after the FET driving source conductance is calculated from the input coils design.

The loaded Q of the input circuit must be high enough to ensure that the frequency response of the RF stage will satisfy the selectivity requirement of 45-dB attenuation at 21.4 MHz *off resonance*. The frequency response of a single-tuned transformer is given by

$$\rho(\text{dB}) = 20 \log_{10}\left[1 + \left(\frac{Q_L 2\Delta F}{F_0}\right)^2\right]^{1/2}$$

where ρ = attenuation at ΔF Hz off resonance and F_0 = resonant frequency.

Attenuation values calculated from this equation at 5.35 MHz and 21.4 MHz off resonance from a center frequency of 100 MHz are given in Fig. 2-121 for several values of loaded Q. (The 5.35-MHz values are included because they are useful in predicting $F_0 + 1/2$ IF rejection.)

The attenuation at ΔF Hz off resonance for *two* single-tuned transformers is found simply by summing the two appropriate attenuation values. The attenuations at the image frequency from Fig. 2-121 are also presented in graphic form in Fig. 2-122 to point out how increasing values of loaded Q rapidly reach a point of diminishing returns. As indicated earlier, the output Q_L will be somewhat less than 68, but still in the neighborhood of 68.

Q_L	Δf (MHz)	Attenuation, ρ (dB)
10	5.35	3.3
10	21.40	12.9
20	5.35	7.5
20	21.40	18.7
30	5.35	10.5
30	21.40	22.2
40	5.35	12.9
40	21.40	24.7
50	5.35	14.7
50	21.40	26.6
60	5.35	16.3
60	21.40	28.2
70	5.35	17.6
70	21.40	29.5
80	5.35	18.7
80	21.40	30.7
90	5.35	19.7
90	21.40	31.7
100	5.35	20.6
100	21.40	32.6
110	5.35	21.4
110	21.40	33.5

Fig. 2-121 Attenuation versus frequency for a single-tuned transformer, center frequency = 100 MHz (Courtesy, Texas Instruments Incorporated)

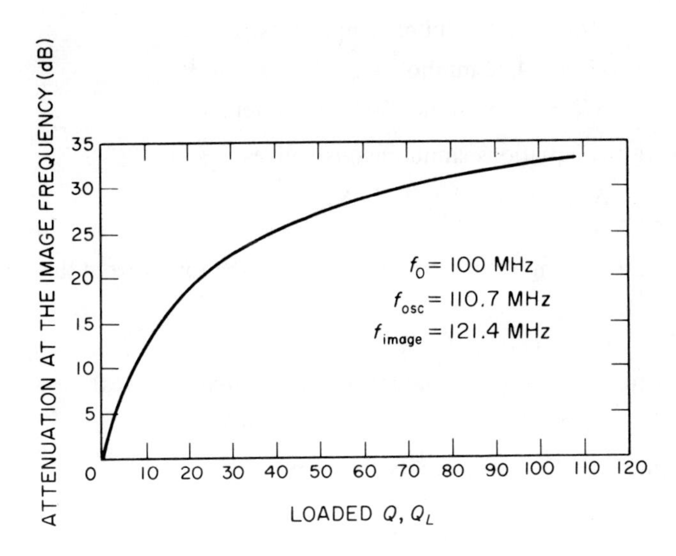

Fig. 2-122 Attenuation at the image frequency as a function of loaded Q for a single-tuned transformer (Courtesy, Texas Instruments Incorporated)

From Fig. 2-121, a loaded Q of 60 will provide 28.2-dB attenuation at the image frequency. Choosing an input loaded $Q = 20$ adds an additional 18.7-dB attenuation at the image frequency so that the selectivity requirement will be satisfied.

The input conductance of the FET is found from:

$$G_{\mathrm{IN}} \approx g_{11} - \mathrm{Re}\left[\frac{y_{12}y_{21}}{g_{22} + G_L}\right] = G_{\mathrm{IN}} \approx 4.22 \text{ mmho}$$

$$R_{\mathrm{IN}} = \frac{1}{G_{\mathrm{IN}}} = 236$$

To minimize input loss, let the input coil $Q_U = 200$. Next, the tapping ratios on the input coil need to be determined. In Fig. 2-123, all conductances have been referred to the top of the input tank, and the following equations apply:

$$\frac{G_{\mathrm{IN}}^*}{(N_{1A}/N_{2A})^2} = \frac{1}{Q_U X_C} + \frac{G_{\mathrm{IN}}}{(N_{1A}/N_{3A})^2}$$

$$\frac{1}{Q_L X_C} = \frac{1}{Q_U X_C} + \frac{G_{\mathrm{IN}}}{(N_{1A}/N_{3A})^2} + \frac{G_S^*}{(N_{1A}/N_{2A})^2}$$

Already established values for substitution into these equations are:

$G_{\mathrm{IN}}^* = 3.33$ mmho (match antenna admittance)

$Q_L = 20$ (for selectivity)

$X_C = 48.4$ ohms

$$Q_U = 200 \text{ (minimize input loss)}$$
$$G_{IN} = 4.22 \text{ mmho}$$
$$G_S^* = 3.33 \text{ mmho (300-ohm antenna)}$$

Solving the equations simultaneously gives:

$$\frac{N_{1A}}{N_{2A}} = 2.5 \quad \text{and} \quad \frac{N_{1A}}{N_{3A}} = 3.2$$

The FET driving source conductance may be found from this equation:

$$\frac{G_S}{(N_{1A}/N_{3A})^2} = \frac{1}{Q_U X_C} + \frac{G_S^*}{(N_{1A}/N_{2A})^2}$$

Substituting appropriate values into this equation gives:

$$G_S = 6.35 \text{ mmho } (R_S = 158 \text{ ohms})$$

which results in a Stern k stability factor of

$$k = \frac{2(4.95 + 6.35)(0.05 + 0.304)}{0.545 + 0.258} = 10$$

A k factor of 10 produces a highly stable RF stage.

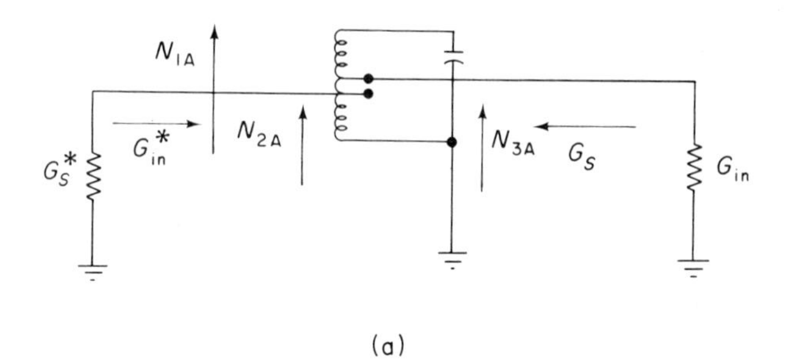

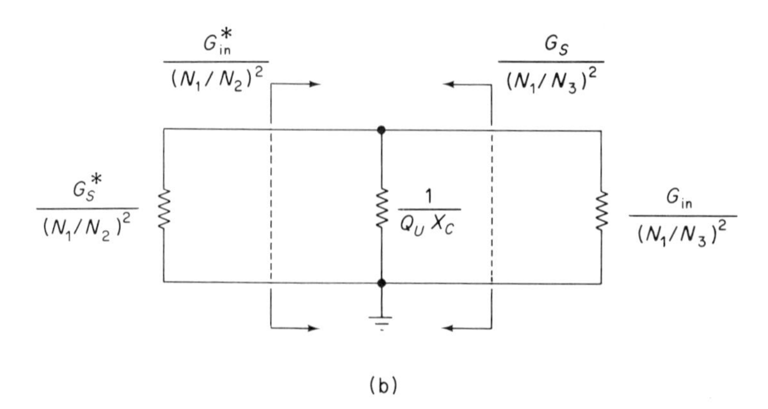

Fig. 2-123 Input coil design (Courtesy, Texas Instruments Incorporated)

Although the optimum driving source impedance for the TIS34 is approximately 500 ohms, the above value of 158 ohms should be satisfactory since the FET noise figure is excellent for a broad range of driving source impedances.

The power loss of the input single-tuned network is given by:

$$\textit{Input power loss} = 10 \log \frac{\left[\dfrac{G_S^*}{(N_{1A}/N_{2A})^2} + \dfrac{G_{\text{IN}}}{(N_{1A}/N_{3A})^2}\right]^2}{\left[\dfrac{G_S^*}{(N_{1A}/N_{2A})^2}\right]\left[\dfrac{G_{\text{IN}}}{(N_{1A}/N_{3A})^2}\right]} + 20 \log \left[\frac{Q_U}{Q_U - Q_L}\right]$$

The first term of the equation represents a mismatch between antenna admittance and the FET input admittance, while the second term represents the insertion loss of the tuned circuit. Since the second term degrades noise figure, the necessity of having a high Q_U to Q_L ratio is apparent. For this particular example,

$$\textit{Input power loss} = 10 \log \left\{\frac{(0.52 + 0.416)^2}{4(0.52)(0.416)}\right\} + 20 \log \frac{200}{200 - 20}$$

Input power loss $\approx$ 1.0 dB.

The FET driving source conductance having been determined, the FET output conductance may be calculated as follows:

$$G_{\text{OUT}} = g_{22} - \text{Re}\left[\frac{y_{12}y_{21}}{G_{11} + G_S}\right]$$

$$G_{\text{OUT}} = 0.027 \text{ mmho} \left(R_{\text{OUT}} = \frac{1}{G_{\text{OUT}}} = 37.0 \text{ k}\Omega\right)$$

A more exact calculation for output tank loaded Q can now be made by using this equation:

$$\frac{1}{Q_L X_C} = \frac{1}{Q_U X_C} + \frac{G_{\text{OUT}}}{(N_{1B}/N_{2B})^2} + \frac{G_L^*}{(N_{1B}/N_{3B})^2}$$

The appropriate values for this equation are:

$$X_C = 48.4 \text{ ohm}$$

$$Q_U = 200$$

$$G_{\text{OUT}} = 0.027 \text{ mmho}$$

$$G_L^* = 4.00 \text{ mmho}$$

$$\frac{N_{1B}}{N_{2B}} = 1.0$$

$$\frac{N_{1B}}{N_{3B}} = 5.0$$

so that $Q_{L(\text{output})} = 62.5$, thus confirming the initial assumption that the FET output impedance would not appreciably load the output coil.

The output power loss which is not included in the general power gain expression of Sec. 2-25-3 cannot be calculated in the same fashion as the input loss. The mismatch loss which is accounted for in the power gain expression is the mismatch between the FET's output conductance and the *total* load seen by the FET.

The mismatch term in the Input Power Loss equation is the mismatch only between the stage driving source conductance and the FET input conductance. Thus, if the general power gain expression is to be used in calculating the overall RF stage power gain, the additional output power loss that must be considered is not a simple function of Q_U and Q_L, but instead is the ratio of power delivered to the total FET load G_L, to the power delivered to actual load G_L^*.

$$\text{Output loss} = \frac{\text{Power delivered to } G_L}{\text{Power into } G_L^*}$$

or

$$\text{Output loss (dB) } 10 \log \frac{\left[\dfrac{G_L}{(N_{1B}/N_{2B})^2}\right]}{\left[\dfrac{G_L^*}{(N_{1B}/N_{3B})^2}\right]}$$

Using this equation, the output loss is

$$10 \log \frac{(0.304/1)}{(4.00/25)} \qquad \text{Output loss} = 2.8 \text{ dB}$$

Using the general power gain expression,

$$G_P \approx 13.5 \text{ (or 11.3 dB)}$$

The overall RF stage power gain then is

$$PG = G_P - \text{input power loss} - \text{output loss}$$

$$PG = 11.3 - 1.0 - 2.8$$

$$PG = 7.5 \text{ dB}$$

The overall voltage gain of the RF stage in dB is

$$VG(\text{dB}) = PG - 10 \log \frac{R_{\text{IN}}^*}{R_L^*} = 7.5 - 10 \log \frac{300}{200}$$

$$VG = 5.7 \text{ dB}$$

Summarizing the coil design:

Input Coil

$$Q_U = 200$$

$$Q_L = 20$$

$$L = 0.077\ \mu\text{H}$$

$$\frac{N_{1A}}{N_{2A}} = 2.5$$

$$\frac{N_{1A}}{N_{3A}} = 3.2$$

Output Coil

$$Q_U = 200$$

$$Q_L = 62.6$$

$$L = 0.077\ \mu\text{H}$$

$$\frac{N_{1B}}{N_{2B}} = 1.0$$

$$\frac{N_{1B}}{N_{3B}} = 5.0$$

Index